New Perspectives in Behavioral Cybersecurity II

As the digital world expands and cyber threats grow more sophisticated, the need for insights from diverse disciplines becomes crucial. Following on from the editor's 2023 title *New Perspectives in Behavioral Cybersecurity*, this book presents studies covering a wide range of the latest topics in cybersecurity—from hybrid intelligence in banking security to the connection between physical and cybersecurity attitudes.

This volume introduces innovative perspectives from countries as varied as Brazil, Bulgaria, Cameroon, and the Philippines, among others, reflecting the global nature of cyber challenges. *New Approaches in Behavioral Cybersecurity II: Human Behavior for Business, Profiling, Linguistics, and Voting* brings together international perspectives that explore how human behavior intersects with cybersecurity. The chapters highlight the integration of behavioral sciences such as psychology, economics, and sociology with traditional cybersecurity approaches. Contributors examine linguistic differences in cyberattacks, explore the impact of personality on hacking behavior, and provide insights into ethical practices in the digital age. The reader will be able to take a different and international look at the complex and evolving world of cybersecurity.

An ideal read for cybersecurity professionals, human factors practitioners, academics, and students, this book will help readers broaden their understanding of how human behavior influences cyber defenses.

Wayne Patterson is a retired professor of computer science at Howard University. In 1993, he was appointed Vice President for Research and Professional and Community Services, and Dean of the Graduate School at the College of Charleston, South Carolina. His other service to the graduate community in the United States has included being elected to the Presidency of the Conference of Southern Graduate Schools, and also to the Board of Directors of the Council of Graduate Schools. Dr. Patterson has published more than 50 scholarly articles primarily related to cybersecurity, one of the earliest textbooks in cybersecurity, *Mathematical Cryptology* (1986), and recently *New Perspectives in Behavioral Cybersecurity* (CRC Press, 2024).

He received degrees from the University of Toronto (BSc and MSc in mathematics), University of New Brunswick (MSc in computer science), and the University of Michigan (PhD in mathematics). He also held Post-Doctoral appointments at Princeton University and the University of California—Berkeley. He has recently been honored to have been named to the "Hall of Honour" at his undergraduate University, St. Francis Xavier University in Nova Scotia, Canada, and also by having a Graduate Fellowship named after him at the College of Charleston, in recognition of his being the Founding Dean of the Graduate School there.

Front Cover

New Perspectives in Behavioral Cybersecurity II

Photographs:

Top Row, L to R: Ann N. Amah, University of Buea, Cameroon; Carlos Azzoni, University of Sao Paulo, Brazil; Jeremy Blackstone, Howard University, USA; Rozalina Dimova, Technical University of Varna, Bulgaria; Tihomir Dovramadjiev, Technical University of Varna, Bulgaria.

Second Row, L to R: Rusko Filchev, Technical University of Varna, Bulgaria; Diana Florea, Lucian Blaga University of Sibiu, Romania; Kaido Kikkas, Tallinn University of Technology, Estonia; Tova Lane, Lane Behavioral Consulting, USA/Israel; Birgy Lorenz, Tallinn University of Technology, Estonia.

Third Row, L to R: Rosangela Malachias, Rio de Janeiro State University, Brazil; Badrouzamani Mana, University of Buea, Cameroon; Augustine Orgah, Xavier University of Louisiana, USA; Dyaisha Orgah, Tulane University, USA; Wayne Patterson, Patterson and Associates, Canada/USA.

Fourth Row, L to R: Ariel Isaac Posada Barrera, Universidad Popular Autonome del Estado de Puebla, Mexico; Aryn Pyke, United States Military Academy (West Point), USA; Laura Margarita Rodriguez Peralta, Universidad Popular Autonome del Estado de Puebla, Mexico; Erick Rosete Beas, Universidad Popular Autonome del Estado de Puebla, Mexico; Michael Ekonde Sone, University of Buea, Cameroon.

Fifth Row, L to R: Jeremiah D. Still, Old Dominion University, USA; Mary L. Still, Old Dominion University, USA; Sneha Sudhakaran, Florida Institute of Technology, USA; Leonid Vagulin, Tallinn University of Technology, Estonia; William Emmanuel S. Yu, Ateneo de Manila University, Philippines.

New Perspectives in Behavioral Cybersecurity II

Human Behavior for Business, Profiling, Linguistics and Voting

Edited by
Wayne Patterson

CRC Press
Taylor & Francis Group
Boca Raton London New York

CRC Press is an imprint of the
Taylor & Francis Group, an **informa** business

Designed cover image: Wayne Patterson

First edition published 2026
by CRC Press
2385 NW Executive Center Drive, Suite 320, Boca Raton FL 33431

and by CRC Press
4 Park Square, Milton Park, Abingdon, Oxon, OX14 4RN

CRC Press is an imprint of Taylor & Francis Group, LLC

ISBN: 978-1-032-98352-3 (hbk)
ISBN: 978-1-032-98528-2 (pbk)
ISBN: 978-1-003-59914-2 (ebk)

DOI: 10.1201/9781003599142

Typeset in Times
by Newgen Publishing UK

Contents

SECTION I   Perspectives on Behavioral Science Approaches in Cybersecurity

SECTION II   Perspectives on Business Approaches in Cybersecurity

Foreword

I'm so glad Wayne Patterson asked me to write this foreword because it's the perfect place to thank him for his incredible impact on me and my career. In an interesting parallel, Wayne, while he was a doctoral student in math at the University of Michigan at Ann Arbor, met a well-known mathematician named Peter Hilton. Hilton had previously worked at Bletchley Park with Alan Turing, attempting to crack the Enigma code. Hilton's discussion of his work at Bletchley Park in a lecture in Ann Arbor inspired Wayne's interest in cryptography and, ultimately, his broader interest in cybersecurity.

Wayne is my Peter Hilton, but his impact on me was much greater than simply choosing a field of study. I was a self-taught programmer and teenage hacker turned computer science student when I met Wayne at the University of New Orleans, where he was a faculty member and then Vice Chancellor for Research. I was positive I would earn my undergraduate degree and then return to programming. Wayne "forced" me to do undergraduate research in cryptography and, in doing so, completely changed my life. My next stop was Ohio State, where I earned my doctorate, and I've been hacking in academia ever since. Thank you, Wayne.

But on to the book, this new volume of *New Perspectives in Behavioral Cybersecurity* continues in the footsteps of its predecessor by addressing an important aspect of cybersecurity that's often underrepresented—the human factor. Technical cybersecurity, the arena where complex malware is reverse-engineered and sophisticated monitoring systems thwart cyberattacks, is crucially important. But human factors, which don't get as much airplay, are equally critical in defending computer systems, because most costs related to cyberattacks are the direct result of human error.

In addition to complex technical attacks, bad actors rely on social engineering and deception to exploit systems, preying on cognitive biases, overload, and social norms, such as respect for authority, reciprocity, and politeness. Insider threats are also a major factor in securing systems, as insiders have more access to critical systems and security controls. Behavioral cybersecurity adds tools from psychology, sociology, and behavioral science to our defensive toolkit.

In this volume, Wayne has compiled contributions from a number of researchers in behavioral cybersecurity. They compare our attitudes on physical security vs. cybersecurity, survey password practices, develop novel authentication schemes that specifically address human cognitive limits, profile the personalities of cybercriminals, and more.

If you're curious about behavioral cybersecurity, this book is a great place to start. If you're a researcher, this field is still wide open—jump in and help make us all safer.

Golden George Richard III
Professor of Computer Science
Louisiana State University

Preface

Since the introduction and proliferation of the Internet, problems involved with maintaining cybersecurity have grown exponentially and evolved into many forms of exploitation.

Yet, cybersecurity has had far too little study and research. Virtually all of the research that has taken place in cybersecurity over many years has been done by those with computer science, electrical engineering, and mathematics backgrounds.

However, many cybersecurity researchers have come to realize that to gain a full understanding of how to protect a cyberenvironment requires not only the knowledge of those researchers in computer science, engineering, and mathematics, but those who have a deeper understanding of human behavior: researchers with expertise in the various branches of behavioral science, such as psychology, behavioral economics, and other aspects of brain science.

The contributors, arising from several disciplines, have attempted in the past few years to understand the contributions that distinct approaches to cybersecurity problems can benefit from this interdisciplinary approach that we have tended to call "behavioral cybersecurity".

Our decision to create this book arose from a similar perspective. Our training is in computer science and psychology, among other fields, and we have observed, as have many other scholars interested in cybersecurity, that the problems we try to study in cybersecurity require not only most of the approaches in computer science, but more and more an understanding of motivation, personality, and other behavioral approaches in order to understand cyberattacks and create cyberdefenses.

Several years ago, we published *New Approaches in Behavioral Cybersecurity* (CRC Press, 2024), a book that would attempt to bring together a diverse set of scholars, as we indicated above, who had developed areas of research even though they were from many nationalities, countries, and cultures; and a wide distribution of ages and genders. And that decision helped in developing our earlier book, which we published as *New Approaches in Behavioral Cybersecurity* (I) and now informs this new book for *New Approaches in Behavioral Cybersecurity II.*

NEW SCHOLARS IN BEHAVIORAL CYBERSECURITY II

As with any new approaches to solving problems when they require knowledge and practice from distinct research fields, there are few people with knowledge of the widely separate disciplines, so it requires an opportunity for persons interested in either field to gain some knowledge of the other. We have attempted to provide such a bridge in this book that, continuing our previous one, we have entitled *New Approaches in Behavioral Cybersecurity II,* continuing in a similar approach to our efforts in the previous book. In the previous book, we were able to publish the innovative work of such scholars, from more than a dozen countries, and with a significant percentage of both female and male authors. Now, we are expanding on the previous research efforts with this book, that we have entitled *New Approaches in Behavioral Cybersecurity II*.

In this new book, we have tried to provide an introductory approach in both psychology and cybersecurity, and as we have tried to address some of these key problem areas, we have also introduced topics from other related fields such as linguistics, banking, network analysis, political science, criminal justice, mathematics, and behavioral economics.

We entered the computer era over 80 years ago. For over three-quarters of that time, we could largely ignore the threats that we now refer to as cyberattacks. There were many reasons for this. There was considerable research done going back to the 1970s about approaches to penetrate computer environments, but there were several other factors that prevented the widespread development of cyberattacks. Thus, the scholarship into the defense (and attack) of computing environments remained of interest to a relatively small number of researchers, primarily from mathematics and computer science.

Beginning in the 1980s, a number of new factors came into play. First among these was the development of the personal computer, which now allowed for many millions of new users with their own individual access to computing power. Following closely on that development was the expansion of network computing, which evolved into the openly available Internet. Now, and with the development of tools such as browsers to make the Internet far more useful to the world's community, the environment was set for the rapid expansion of cyberattacks, both in number and in kind, so the challenge for cybersecurity researchers over a very short period of time became a major concern to the computing industry.

The world of computer science was thus faced with the dilemma of having to adapt to changing levels of expertise in a very short period of time. The editor of this book began his own research in 1980, in the infancy of what we now call cybersecurity, even before the widespread development of the personal computer and the Internet.

CONTRIBUTIONS FROM DIVERSE DISCIPLINES

In this book, we have tried to achieve several objectives in addressing a number of the issues involving cybersecurity threats in many different types of environment. In addition, we have come to understand that the entire cyber-environment in which most of us live and participate, has threats that can emerge from many directions and at many levels of sophistication.

As interest in a broad range of approaches to providing greater opportunities for defending and securing one's computer environment has grown, we find that there is a rapidly growing field of scholars, emerging both from fields of computer science and cybersecurity, and also from a range of behavioral sciences including psychology, criminal justice, law, sociology and other human behavior-related disciplines.

In addition, we have noted and presented in this book, contributions that illustrate that this aspect of cybersecurity, once thought to be only studied in a few highly developed countries, now has serious scholarship in a wide range of countries all over the world. This book features contributions from, for example, Brazil, Bulgaria, Cameroon, Canada, Estonia, Mexico, the Philippines, Romania, as well as the United States.

Scholarship in this field has interested many new young researchers, as indicated by many of the contributors to this book, not only coming from diverse national environments, but also illustrating that this aspect of cyber security has been an interest in encouraging scholarship from a significant number of women scholars—around 45% of our contributing authors. This is notable in that the last decades of scholarship in both computer science and in the field of cyber security began to evolve, had very few contributions from women scholars.

Of the research articles presented in this book, we have somewhat arbitrarily divided them into five groups, as follows:

SECTION I: PERSPECTIVES ON BEHAVIORAL SCIENCE APPROACHES IN CYBERSECURITY

The first section addresses the work of a set of authors who address perspectives on protecting business and enterprise from attacks designed to jeopardize their occupation.

The first chapter begins with a contribution from a team from the Universidad Popular Autonome del Estado de Puebla (UPAEP), Puebla, Mexico (Ariel Barrera, Laura Margarita Rodriguez Peralta, and Alan David Jimenez), "Integrating Behavioral Sciences and Deep Learning in Network Behavior Analysis for Threat Detection". They use a technique called Network Behavior Analysis (NBA) which attempts to use behavioral science and machine learning to detect profiles of potential attack sources, and thus assist network administrators to improve audit processes.

Next, it is followed by a contribution from Jeremiah Still and Mary Still of the Old Dominion University in Virginia, "Cognition-Inspired Design in Mind: Demonstrations in Graphical Authentication". Their contribution addresses the problem of having many distinct passwords for users in a system. This technique, "graphical authentication" aims to detect profiles of potential attack sources. Deep Learning algorithms are applied to predict potential threats. This section ends with a presentation given by Wayne Patterson to a large senior citizen's residence in Washington, DC, advising on methods for the audience to take safeguards against cyberattacks on their personal computer systems.

SECTION II: PERSPECTIVES ON BUSINESS APPROACHES IN CYBERSECURITY

This section will address cybersecurity problems in three areas. From the banking sector, a team from the Technical University of Varna in Bulgaria (Rusko Filchev, Tihomir Dovramadjiev, and Rozalina Dimova) contribute a chapter "Hybrid Intelligence in Cybersecurity Banking". The authors address sustainability in banking, especially the importance of the human factor in protecting digital assets. The convergence of AI and human intelligence fosters sustainability. The authors argue for a comprehensive approach. They explore how the hybrid intelligence (HI) can create resilient security systems.

That is followed by a chapter addressing the development of an intrusion tolerant system using aspects of computer and human behavior: "Efficient Intrusion Tolerant System Based on Machine Learning and Human Behavior". Michael Ekonde Sone, Ann Amah, and Badrouzamani Mana of the University of Buea in Cameroon develop an intrusion tolerant system to determine model parameters, using Markov decision processes, for transition probabilities and control. They implement an intrusion tolerant system using a hidden Markov model. This system has a better prediction for known attacks.

Section III: Perspectives on Profiling Approaches in Cybersecurity

Tova Lane (of Lane and Associates) integrates four disciplines: psychology, neuroscience, criminology, and cyberpsychology to profile a criminal. She develops a model based on the European Commission for profiling the criminal. Dr. Lane also addresses personality characteristics of hackers.

Aryn Pyke of the United States Naval Academy addresses physical security attitudes and cybersecurity attitudes in her chapter, "Cybersecurity Is Not on Maslow's Hierarchy: Implications of the Difference between Users' Physical Security and Cybersecurity Attitudes and Behaviors". She compares analysis of users' physical security and cybersecurity attitudes and behaviors with Maslow's hierarchy. She analyzes attitudes and practices on a self-reporting scale, and concludes that practices correlate negatively with users' computer trust, and positively with their computer expertise and physical security practices.

Section IV   Perspectives on Linguistic Approaches in Cybersecurity

A team of seven contributors from five countries, led by Wayne Patterson, will demonstrate a technique in "Using Language Translation Software to Detect and Classify Cyberattacks from Suspected Cyberattackers from Specific Countries and Languages". This team includes Carlos Azzoni and Rosangela Malachios (Brazil); Diana Florea (Romania); Kaido Kikkas, Birgy Lorenz, and Leonid Vagulin (Estonia); and William Yu (Philippines). The countries in question were initially studied in the *World Cybersecurity Index* at Oxford.

This chapter is followed by a study of a different set of countries investigated in "Isolating Key Phrases to Identify Ransomware Attackers" by Jeremy Blackstone of Howard University in Washington. His approach uses linguistic analysis of ransomware messages to ascertain the adversary's language of origin. He isolates use of these phrases to rank the languages from easiest to most difficult.

Section V   Perspectives on Voting Approaches in Cybersecurity

William Emmanuel S. Yu, of the Ateneo de Manila University in the Philippines, contributes "Protecting Democracy in an Increasingly Digital World: A Philippine e-Voting Story". As the world gets smaller with the use of technology, people expect more and more from it. With advancements in technology, we now perform our transactions anytime, anywhere, and with any device. No more long lines. No more banking hours (24 × 7 banking is now the norm). No more having to carve out a large part of one's day to do banking. On the flip side, we also see more victims of phishing and account takeover attacks. At the end of 2022, digital payments already account for 42.1% of all Philippine financial transactions.

A team from three universities in Louisiana and Florida, led by Austin Orgah (Xavier University of Louisiana), with Sneha Sudhakaran (Florida Institute of Technology) and Dyaisha Orgah (Tulane University), address "Passwords: An Empirical Study of Modern Behaviors in the Social Media Era".

Acknowledgments

The editor is extremely grateful for the learned contributions from the 25 coauthors, female and male, from so many different countries, who have provided important parts of this volume.

Contributors

Ann N. Amah (amahpraises@gmail.com.) is a PhD student in Telecommunications and Networks Engineering at the University of Buea, Cameroon, holding a master's degree in the same field.

Carlos Azzoni, is a former Dean of the Faculty of Economics, Administration and Accounting of the University of Sao Paulo, Brazil. His field of expertise is regional and urban economics. He was Visiting Professor at the University of Illinois and the Ohio State University. He is Fellow of the Regional Science Association International.

Jeremy Blackstone is Assistant Professor at Howard University in the Electrical Engineering and Computer Science Department. He is a two-time alumnus of Howard University receiving his Bachelor's and Master's degree in Systems and Computer Science at Howard University and his PhD in Computer Science and Engineering from University of California, San Diego (UCSD). He performs research in behavioral security, hardware security, network security, and artificial intelligence working on projects developing solutions for side channel attack mitigation, ransomware mitigation, intrusion detection, vulnerability prediction modeling, and process scheduling optimization. He serves as the co-director of the Matthew Henson Fellowship program which provides 12 months of funding for undergraduate students at Howard University to work on research projects while receiving mentorship from researchers at both UCSD and Howard University. He also teaches courses on C++ syntax and data structures (Computer Science I, II, III), security research (Intro to Cybersecurity II, Computer Security II), and Robotics Autonomous Navigation and received a Student Choice Professor Award for academic year 2023–2024 from the Howard University College of Engineering and Architecture Student Council for the impact of his teaching and mentorship and contributions.

Rozalina Dimova has many years of activity at the Technical University of Varna, Bulgaria, at the position of Rector, as well as Vice-Rector and Dean.

Dimova has research interests in cybersecurity, ergonomics, and human factors, and is currently on the board of directors of the Bulgarian Association of Ergonomics and Human Factors (BAEHF); develops academic and educational activities related to ICT publications; and has participated either as project leader or member of scientific teams in 5 European and more than 40 national-funded research projects.

Dimova is a member of European Society for Engineering Education (SEFI), IEEE, a member of Management Board of the Scientific and Technical Union-Varna, Bulgaria, and a member of Management Board of Union of Scientists in Bulgaria.

Tihomir Dovramadjiev carries out and develops professional activity related to academic, educational, and scientific activities related to the Technical University—Varna (TUV, Bulgaria) and the Bulgarian Association of Ergonomics and Human Factors (BAEHF). He received the PhD in "Ergonomics and industrial design" (TUV/2012). He has experience as a lecturer for more than 15 years at Industrial Design Department (TUV). He is the author of the book *Advanced technologies in Design*, TUV, pp. 228, ISBN: 978-954-20-0771-5, 2017. He has participated in over 90 scientific publications (including Springer and Elsevier). He actively participates in scientific conferences IHSED, IHSI, AHFE, and IHIET as a scientist, reviewer, and organizer. He is on the editorial board of the *IETI Transactions on Ergonomics and Safety (TES)*, *IETI Transactions on Engineering Research and Practice (TERP)*, and *American Journal of Management Science and Engineering (AJMSE)*; and topical advisory member of MDPI Mathematic and other.

Rusko Filchev is a long-term bank manager (Societe Generale Expressbank, DSK, and others). He is responsible for information and telecommunication security of the bank office. Its activities cover innovations in cybersecurity and monitoring of high-tech processes related to financial institutions. He organizes and is responsible for the implementation of the goals and tasks approved by the business plan and budget of the branch. He is also responsible for the good development of the business, makes contacts to attract new clients in the office, maintains relationships with current clients, and researches opportunities to offer new products and services. He has a university Master degree in Economics. His competence covers as well as performs activities in the following areas: public sector economics; pension insurance funds; public sector audit; health insurance systems; European Union finance; research; government debt management; tax process; fiscal policy and mechanisms; and others.

Diana Florea is Lecturer of Information Management in the Department of Romance Studies of the "Lucian Blaga" University of Sibiu, Romania. Her current interests are in cultural histories of reading, cultural and knowledge management, and information theories.

Kaido Kikkas (b. 1969) is the Associate Professor of Information Society and Cyberculture at the IT College of Tallinn University of Technology (Tallinn, Estonia). His focus in research and teaching has been a wide range of ethical and social issues of IT (information society, licensing, hacker ethics, etc.), "softer" topics in cybersecurity, and different ways of learning. In his free time, he plays keyboards and guitars, reads books, practices martial arts, cooks, and (during the season) rides Uncle Hawk, his Indian motorcycle.

Tova Lane is a psychologist in private practice with a specialization in profiling cybercriminals. With experience in both clinical psychology and cybersecurity, Dr. Lane consults on cybersecurity issues, bridging the gap between psychological insights and technological defenses. Dr. Lane's unique expertise provides valuable perspectives on the motivations and behaviors of cybercriminals.

Birgy Lorenz has extensive expertise as a cybersecurity awareness trainer, specializing in security behavior, common people cyber hygiene, and teaching parents, teachers, and kids about digital safety. She oversees numerous awareness initiatives in which she collaborates with educators and young hackers in the cybersecurity field. For example, she has created cybersecurity curriculum for secondary schools and an exercise webpage for aspiring cybersecurity talent. Informatic Europe 2018 acknowledged her project CyberOlympic (competitions, training camps), and Estonia 2022 awarded her a Science Populator Award. She was also awarded Teacher of the Year in Estonia in 2010, 2009, and 2017. She currently serves as the Vice Dean of BA studies at Tallinn University of Technology's (TalTech) IT School.

Rosangela Malachias, is Associate Professor at Rio de Janeiro State University/College of Education from Baixada Fluminense (Brazil); Professor of the Graduate Program in Communication (PPGCom-UERJ); Coordinator and President of the Permanent Committee of Self-Declaration Validation (CoopVA/CPVA-UERJ) for Affirmative Action (quotas); Full member of PPGEDUC—Graduate Program of the Federal and Rural University from Rio de Janeiro; and Coordinator of AFRODIÁSPORAS Research Center on Black Women, Visual Culture, Educommunication and Policies in Urban Peripheries.

Badrouzamani Mana is a graduate student at the University of Buea, Cameroon, pursuing a Master's degree in Software Engineering. He holds a Bachelor's degree from the same field. He can be reached at bmana917@gmail.co

Augustine Orgah is a computer science instructor and Program Manager for Xavier University of Louisiana's Physics and Computer Science department where he teaches computer science and cybersecurity courses while nurturing the budding computing students in their development. A security researcher and enthusiast, he enjoys teaching and is committed to building a cybersecurity pipeline at Xavier University. Contributing to the book is part of that journey toward improving cybersecurity education, recruitment, and retention especially among minorities. His main research area is information assurance/cybersecurity, specifically memory forensics and malware analysis. In his spare time, he enjoys soccer, reading, and watching shows about nature.

Dyaisha Orgah currently holds the position of IT Application Manager III at Tulane University. Before her tenure at Tulane, Dyaisha amassed eight years of experience as an IT Business Systems Analyst at Louisiana's largest media company. Her professional journey has been enriched by engagements across diverse sectors, including healthcare, casino gaming, funeral services, and professional sports.

Throughout her career, Dyaisha has demonstrated a steadfast commitment to enhancing operational efficiencies and facilitating organizational effectiveness in alignment with business technology objectives. Dyaisha holds a Bachelor's degree in Organizational Information Technology and a Master's degree in IT Management from Tulane University, underscoring her academic prowess and dedication to advancing her field of expertise. Her passion lies in optimizing business processes to empower organizations to achieve their strategic goals.

Ariel Isaac Posada Barrera is currently pursuing a PhD in Information Technology at UPAEP, Puebla. He completed an academic exchange program at Universidad del Rosario. He holds a degree in Engineering in Computational Technologies from Tecnológico de Monterrey and a Master's degree in Information Technology Management from Universidad Tecmilenio. He has professional experience in software engineering and development. His certifications include Digital Forensics Essentials (DFE) v1 from EC-Council, Data Science and Data Engineering from Data Science Dojo, and Senior Data Scientist from Tecnológico de Monterrey. Currently, he is involved in research projects on cybersecurity and indoor air quality analysis.

Aryn Pyke is Cognitive Cyber Research Scientist at the Army Cyber Institute and Associate Professor in the Engineering Psychology Program at West Point. Her interest in the human-in-the-loop in cybersecurity stems from an interdisciplinary background, including a BASc and MASC in electrical and computer engineering and a PhD in cognitive science. Her current research interests include human–computer interaction, STEM education innovations, cybersecurity situational awareness, talent management (e.g., assessment and training innovations), and workflow characterization and augmentation.

Laura Margarita Rodríguez Peralta holds a Bachelor's degree in Computer Science from BUAP and a Master's in Computer Science from UDLAP (Mexico). She obtained her DEA in Telecommunications and Systems from Joseph Fourier University (France) and her PhD in Computer Science and Telecommunications from INPT (France). She has been a professor and researcher at the University of Madeira (Portugal), the University of Salvador (Brazil), and UPAEP (Mexico), where she directed the Faculty of Information Technologies. She has published extensively and participates in research projects, focusing on IoT, cybersecurity, and computer networks.

Erick Rosete Beas holds a Bachelor's degree in Mechatronics Engineering from CETYS University, Mexico. He earned a Master's degree in Computer Science with a specialization in Artificial Intelligence from the University of Freiburg, Germany. Currently, he is pursuing a PhD in Information Technology at UPAEP, Puebla. Erick has contributed to various research projects and publications, with a focus on robotics, machine learning, deep learning, and reinforcement learning

Michael Ekonde Sone is Associate Professor in Telecommunications and Networks Security. He is currently the Deputy Vice-chancellor at the University of Buea, Cameroon. He has published extensively in peer-reviewed journals and contributed book chapters in some books in cybersecurity notably *New Perspectives in Behavioral Cybersecurity* by Wayne Patterson and *Computer and Network Security* by J. Sen. He can be reached at michael.sone@ubuea.cm and ekonde.sone@gmail.com

Jeremiah D. **Still** is Associate Professor at Old Dominion University. He earned a PhD in Human-Computer Interaction from Iowa State University in 2009. Dr. Still is Fellow of the Psychonomic Society and received the Earl Alluisi Award from APA Division 21, which recognizes his outstanding achievements as an applied experimental/engineering psychologist. His *Psychology of Design laboratory* explores the relationship between human cognition and technology. Specifically, he focuses on applied visual attention, human-centered cybersecurity, and usability measurement.

Mary L. **Still** is Assistant Professor at Old Dominion University. She earned a PhD in Experimental Psychology from Iowa State University in 2009. Research in Dr. Still's Building on Intuition, Knowledge, and Experience (BIKE) laboratory is focused on understanding how knowledge from previous experience and interpretation of real-time environmental information contribute to attributions, actions, and decisions. In this context, her work has examined how to create intuitive interface interactions while considering the capabilities of human memory and attention.

Sneha Sudhakaran is a distinguished scholar and cybersecurity expert currently serving as Assistant Professor in the Department of Computer Science at the Florida Institute of Technology. Dr. Sudhakaran's research interests encompass various cybersecurity topics, including android security, application security, host security, cyber forensics, and blockchain. As a cyber forensic researcher, she aims to create more profound knowledge for cyber forensic courses for students at the Florida Institute of Technology, which would help students get better jobs. Her certifications further validate her expertise as a Certified Ethical Hacker (CEH) and a Certified Hacking and Forensic Investigator (CHFI). Dr. Sudhakaran's dedication to advancing cybersecurity education and practice is evident in her ongoing research. Her work not only contributes to academic knowledge but also provides practical solutions to real-world cybersecurity challenges, thereby fostering a more secure digital environment.

Leonid Vagulin was born in rural parts of Estonia. He was for a long time a seaman. Now he works in IT for a small Estonian software company. He quite enjoys the tech world and aims to make it a little bit safer.

William Emmanuel S. Yu is Chief Technology Officer responsible for global technology service fulfillment at Novare Technologies, a diversified technology systems integrator, working on various projects involving digital transformation, elastic infrastructure, and end-to-end AGILE DevOps in diverse industries from telecommunications, banking and finance, utilities, and sales and distribution. He works with customers, which include the country's top telecommunications, financial institutions, and select organizations abroad, to help delight their customers with next generation telecommunication and Internet technology.

Outside his corporate endeavors, he is active in technology advocacy and technology voluntarism. He is a part of Secure Connections, a cybersecurity project of The Asia Foundation—Philippines. He is also Trustee for the PPCRV (citizen's arm for the safe guarding the elections), member of the Philippine's Commission on Elections Advisory Council (CAC), and an active member of Internet CSOs such as the Internet Society (Internet governance and policy), and PH-CERT (Information security).

He serves as a faculty member at the Ateneo de Manila University's Computer Science department, and he is a Certified Information Systems Security Professional (CISSP), Certified Secure Software Lifecycle Professional (CSSLP), Certified in Risk and Information Systems Control (CRISC), and a Certified Information Security Manager (CISM). He holds a PhD and a Master's degree in Computer Science, a Bachelor's degree in Computer Engineering, and a Bachelor's degree in Physics from the Ateneo de Manila University.

Future Directions in Behavioral Cybersecurity

It is believed that in order to counter the clever but malicious behavior of hackers and the sloppy behavior of honest users, cybersecurity professionals (and students) must gain some understanding of motivation, personality, behavior, and other theories that are studied primarily in psychology and other behavioral sciences.

Consequently, by building a behavioral component into a cybersecurity program, it is felt that this curricular need can be addressed. In addition, noting that while only 20% of computer science majors in the United States are women, about 80% of psychology majors are women. It is hoped that this new curriculum, with a behavioral science orientation in the now-popular field of cybersecurity, will induce more women to want to choose this curricular option.

In terms of employment needs in cybersecurity, estimates indicate "more than 209,000 cybersecurity jobs in the U.S. are unfilled, and postings are up 74% over the past five years".

It is believed that the concentration in behavioral cybersecurity will also attract more women students since national statistics show that whereas women are outnumbered by men by approximately 4 to 1 in computer science, almost the reverse is true in psychology.

It has also not escaped our notice that the field of cybersecurity has been less attractive to women. Estimates have shown that even though women are underrepresented in computer science (nationally around 25%), in the computer science specialization of cybersecurity, the participation of women drops to about 10%.

However, with the development of a new path through the behavioral sciences into cybersecurity, we observed that approximately 80% of psychology majors, for example, are female. We hope that this entrée to cybersecurity will encourage more behavioral science students to choose this path, and that computer science, mathematics, and engineering students interested in this area will be more inclined to gain a background in psychology and the behavioral sciences.

Introduction to the Book

New Perspectives in Behavioral Cybersecurity II: Human Behavior for Business, Profiling, Linguistics, and Voting offers directions for readers in areas related to human behavior and cybersecurity by exploring a number of new ideas and directions in this subject, by reporting on research and new techniques in this field. The new approaches explored here have come from scholars with very diverse backgrounds: from their disciplinary backgrounds, from their geographic origins and perspectives, from their relative youth, and as a balance of female and male contributors. We seek to demonstrate an understanding of motivation, personality, and other behavioral approaches in understanding cyberattacks and creating cyberdefenses.

We attempt to show cybersecurity issues and perspectives on behavioral science approaches in cybersecurity issues in

- Business
- Profiling
- Linguistics
- Voting
- and other areas of scholarship.

This title is an ideal read for senior undergraduates, graduate students, faculty, and professionals in fields such as ergonomics, human factors, human–computer interaction, computer engineering, psychology, business, linguistics and political science.

In 2019, we decided to explore a number of questions arising from a relatively new field of scholarship that we decided to call "Behavioral Cybersecurity". In that book (*New Perspectives in Behavioral Cybersecurity*, CRC Press, 2024), we identified a number of questions in the field of Cybersecurity that used both mathematical and engineering approaches as well as those from a behavioral science perspective.

We are also very pleased that we are able to include in this book chapter authors who are active scholars in ten countries throughout the world, in particular:

- Brazil
- Bulgaria
- Cameroon
- Canada
- Estonia
- Israel
- Mexico
- Philippines
- Romania
- United States

Since the publication of that book, we have found a rapidly growing interest in this field, so we decided to assemble a number of new contributions in this area of scholarship.

We believe readers will find this an interesting addition and complement to our previous book. We have grouped these papers into five related themes which are:

Section I   Perspectives on Behavioral Science Approaches in Cybersecurity
Section II   Perspectives on Business Approaches in Cybersecurity
Section III   Perspectives on Profiling Approaches in Cybersecurity
Section IV   Perspectives on Linguistic Approaches in Cybersecurity
Section V   Perspectives on Voting Approaches in Cybersecurity

Section I

Perspectives on Behavioral Science Approaches in Cybersecurity

1 Integrating Behavioral Sciences and Deep Learning in Network Behavior Analysis for Threat Detection

Ariel Isaac Posada Barrera, Laura Margarita Rodríguez Peralta, and Erick Rosete Beas

1.1 INTRODUCTION

Classifying user traffic in networks using machine learning techniques is crucial for enhancing security and network management. Detecting not only the type of traffic but also recognizing user types based on statistics is important, as individual traffic classification can be biased by misclassifications or human error. The main challenge lies in identifying whether daily network activities are conducted negligently or maliciously.

This approach closely relates to behavioral sciences, as it allows for analyzing and understanding user behavior patterns in the network. By identifying the statistical characteristics of network traffic and correlating them with specific behaviors,

DOI: 10.1201/9781003599142-2

it is possible to develop models that not only detect security threats but also provide deeper insights into user motivations and habits, as described in Refs. [1] and [2]. This knowledge is invaluable for designing more effective and personalized intervention strategies, thereby improving both network security and the understanding of human behavior in digital environments.

Behavioral sciences play a fundamental role in interpreting the data collected through machine learning techniques. By integrating concepts from psychology and sociology, it is possible to identify behavior patterns that indicate potential risks or atypical behaviors. For example, a sudden change in a user's network usage pattern could signal a possible security breach or an intentional deviation from acceptable use policies.

Moreover, this analysis helps identify users who may be acting maliciously or negligently, which is crucial for implementing security policies and training. By better understanding how and why users interact with the network in certain ways, organizations can create more effective training programs and policies that mitigate the risk of dangerous behaviors.

The combination of advanced machine learning techniques with principles from behavioral sciences offers a powerful tool for network management and security. It not only enhances the detection of malicious activities but also provides a more holistic understanding of user dynamics within the network. This, in turn, facilitates the creation of safer and more efficient digital environments where activities can be proactively monitored and managed.

In this study, the dataset from Ref. [3], publicly available in Ref. [4], was used. This dataset includes detailed data on both benign and malicious network traffic. Data transformation was performed using statistical variables such as minimum, maximum, mean, and variance at the IP address level, providing a detailed statistical overview. The data were then converted to binary format, and deep learning models were applied to improve the accuracy of detecting malicious or negligent behavior. The objectives of this study are as follows:

1. Create a dataset that represents the behavior of a network user by considering their network traffic statistics, obtaining a representation that is easy to store and usable in deep learning algorithms for the classification of benign and malicious behavior.
2. Train machine learning and deep learning models to detect users who may pose potential threats to the network, providing network administrators and security personnel with information for inspections or review of certain devices, minimizing the impact of possible incidents, and reducing operational costs.

This text is organized as follows: Literature Review (Section 1.2) lists previous works on network traffic classification, Methodology (Section 1.3) describes the data transformation performed and the machine learning algorithms applied, Lessons Learned (Section 1.4) presents the findings and the interpretation of the results, Conclusions (Section 1.5) summarizes the conclusions drawn from the obtained knowledge, and Future Works (Section 1.6) discusses future work to be carried out with this new dataset, including its potential use in image-based analyses.

TABLE 1.1
Previous Studies

Study	Approach
[5]	Classification using statistics and frequencies
[6]	Classification using an Ethernet interface monitoring tool
[7]	Traffic classification using statistical values
[8]	Traffic analysis for anomalous traffic classification using k-means
[9]	Classification of compromised devices through network traffic balance analysis with Support Vector Machine
[10]	Empirical analysis of datasets, classifying with Random Forests
[11]	Use of AR, ARMA, ARIMA, and SARIMA regression methods for network traffic classification
[12]	Traffic classification using GrAlg and GNN
[13]	Traffic classification using Deep Learning
[14]	Traffic analysis with the unsupervised Gaussian mixture model
[15]	Network traffic classification using time series with LSTM
[16]	Traffic classification with BERT and LSTM
[17]	Classification using CNN
[18]	Proposal for using a combination of XGBoost, LightGBM, and Random Forest models for network traffic classification
[19]	Proposal for a flow analysis tool to determine user activities and whether it is an intrusion
[20]	Traffic classification using G-TIAG

1.2 LITERATURE REVIEW

In this section, previous studies that apply machine learning techniques to classify network traffic are investigated. These studies have utilized a variety of approaches and machine learning algorithms to address the challenges associated with network traffic classification, including the detection of malicious activities and the differentiation of user behavior patterns. As mentioned in Ref. [3], previous research has typically used statistical features obtained from packet captures, resulting in PCAP files. Table 1.1 lists the existing studies along with the applied machine learning techniques.

In this study, the evaluation will be conducted using a new dataset that leverages the presence of more features in the packet capture. This approach differentiates from state-of-the-art work by classifying behavior not at the packet or segment level, but by using the statistical behavior of IPs to represent user behavior.

1.3 METHODOLOGY

The statistics of the dataset were obtained at the IP level, considering that for the scenario, the IPs are static, assigned to each device, and each computer is used by only one user, with each user using only one computer. We listed the IPs corresponding only to personal computers, not including mobile devices within the current scope, and excluding active network devices and servers. This focused approach allowed for a

TABLE 1.2
Classification Models Results

Model	Accuracy	F1 Score
Support Vector Machines	0.7380	0.7053
Random Forest	0.7500	0.7393
Decision Tree	0.6905	0.6718
Gaussian Naive Bayes	0.7261	0.5849
Gradient Boosting	0.7261	0.7114

more accurate representation of user behavior based on the network traffic generated by personal computers.

For the grouped data by IP, the column for the total number of rows was used. Total counts for protocols and source and destination ports were encoded using one-hot encoding. For continuous values, the minimum, maximum, mean, and standard deviation were calculated. This process resulted in a total of 1,246 columns.

Before experimenting with the images, a data cleaning process was performed. Infinite values were replaced with the maximum, negative infinite values with the minimum, and Not-a-Number (NaN) values with the column mean. For labeling, both the IPs described as attackers in the original paper and those marked as suspicious were considered, with only nonattack IPs labeled as normal traffic. The strategy focused on classifying the IPs that require attention.

This resulted in a dataset of 191 elements classified as normal traffic, and 225 items classified as risky, which include both the IPs that conducted attacks and those considered to have suspicious activity.

Using the dataset, we applied the classification models Support Vector Machines [21], Random Forest [22], Decision Tree [23], Gaussian Naive Bayes [24], and Gradient Boosting [25]. The measures Accuracy [26] and F1 Score [27] were applied to evaluate the models' performance, with the results shown in Table 1.2.

Subsequently, these data were converted into grayscale images by scaling the data into values between 0 and 255 and placing them in a matrix. This transformation allows for leveraging convolutional neural networks (CNNs) for prediction and analysis, which are highly effective in image processing tasks. By converting tabular row data into a visual format, we enhance the interpretability of the data, making it easier to identify patterns and anomalies that may not be as apparent in the raw tabular form. Figure 1.1 presents two examples of IP profiles: subfigure (a) displays a critical labeled IP with normal positioning, indicating that it requires immediate supervision due to its critical nature, while subfigure (b) illustrates a normal labeled IP, characterized by normal traffic statistics and lacking any signs of suspicious activity.

The ResNet models proposed in [28], including ResNet18, ResNet50, ResNet101, and ResNet152, were fine-tuned using the generated images with the support of FastAI [29]. The experimental results are detailed in Table 1.3. ResNet, or Residual Networks, is an advanced neural network architecture originally proposed in [28]

(a) (b)

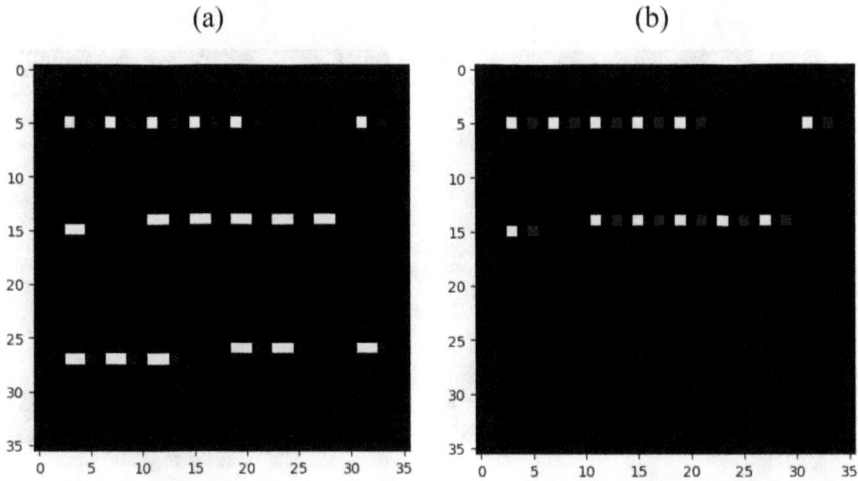

FIGURE 1.1 (a) Critical labeled IP with normal positioning, requiring immediate supervision. (b) Normal labeled IP with normal traffic statistics and no suspicions.

TABLE 1.3
ResNet Models Results

Model	Accuracy	F1 Score
ResNet18	0.7228	0.7228
ResNet50	0.7228	0.7169
ResNet101	0.6385	0.6294
ResNet152	0.6385	0.6372

to enhance the training of deep networks by integrating residual connections. These connections allow information to propagate through layers without alteration, addressing the challenge of vanishing gradients in deep networks. This capability is particularly advantageous for refining the model's ability to analyze intricate and profound features present in generated images, thereby improving prediction accuracy.

Following the method described in [30], known as Vortex Feature Positioning, the pixels were arranged to create spiral-shaped images using Pearson correlation coefficients [31], with the most important features positioned in the center. Figure 1.2 shows both the critical IP and normal IP from Figure 1.1. The images exhibit similar patterns; however, the image representing the critical IP shows higher concentrations of values in the most dependent variables, resulting in larger white shapes. This is due to the traffic having metrics above the average, which is evident in the critical IP image (see Table 1.4).

Based on these results, with ResNet18 identified as the best model achieving an Accuracy of 78.31%, an evaluation was conducted to visually observe the most

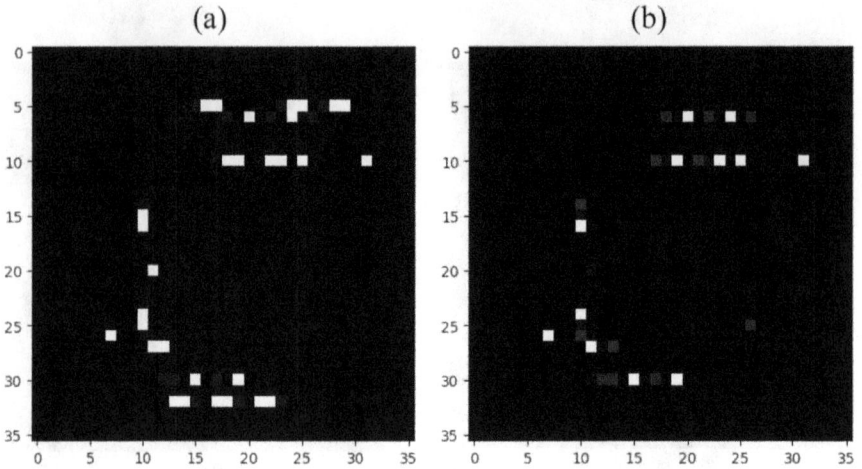

FIGURE 1.2 (a) Critical labeled IP with vortex feature positioning. (b) Normal labeled IP with vortex feature positioning.

TABLE 1.4
ResNet Models Results with Vortex Feature Position

Model	Accuracy	F1 Score
ResNet18	0.7831	0.7828
ResNet50	0.7710	0.7622
ResNet101	0.7469	0.7372
ResNet152	0.7469	0.7415

important variables using the Grad-CAM technique [32]. Figure 1.3 demonstrates the detection of an IP correctly classified as critical and another as normal traffic by this model, highlighting the regions observed by the last convolutional layer. Despite the images appearing very similar, the model effectively distinguishes between these classes.

1.4 LESSONS LEARNED

The images generated from the statistical data and the total observed data were successfully classified using both tabular data machine learning algorithms and image conversion with deep learning. There were notable differences in the metrics achieved by these two approaches.

Among the tabular data algorithms, the Random Forest algorithm emerged as the top performer with an accuracy of 75%. In contrast, within the domain of deep learning, the application of Vortex Feature Positioning resulted in the ResNet18 model achieving the highest accuracy score of 78.31%. This highlights that ResNet18, with

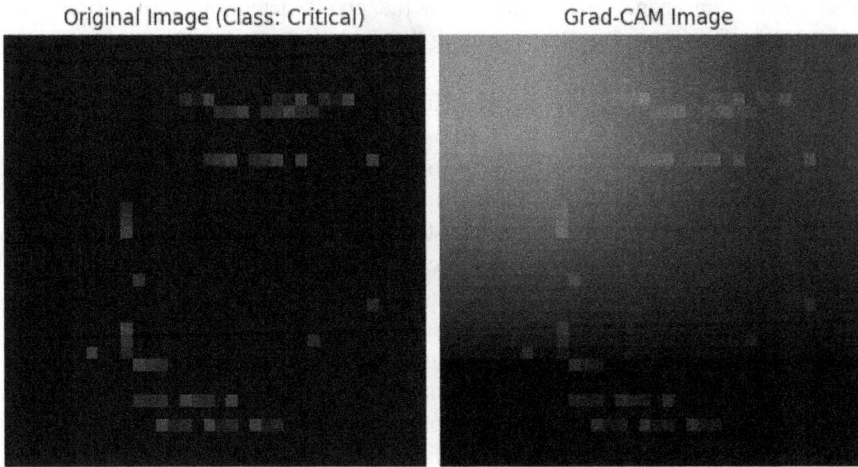

FIGURE 1.3   (a) Grad-CAM image for IP labeled as critical traffic. (b) Grad-CAM image for IP labeled as normal traffic.

its convolutional layers, proved more effective and practical to use compared to its more complex counterparts. This difference in performance could be attributed to the dataset size, as smaller models like ResNet18 often require fewer data points to generalize effectively.

Furthermore, the Grad-CAM technique provided valuable insights by highlighting the consistency in the pixels that determined the critical and normal classifications of an IP. This visualization confirmed that the model was focusing on relevant parts of the images for its classification decisions.

Overall, these findings illustrate the effectiveness of employing both tabular data machine learning algorithms and deep learning image conversion techniques for network traffic classification. The superior performance of the ResNet18 model, particularly when using Vortex Feature Positioning, underscores its potential for practical applications in this domain.

1.5   CONCLUSIONS

The images generated from the statistical and observed total data were successfully classified using both tabular data machine learning algorithms and image conversion with deep learning. Notable differences in metrics were observed between these approaches. The Random Forest algorithm was the best performer for tabular data with an accuracy of 75%. In contrast, Vortex Feature Positioning with the ResNet18 model achieved an accuracy of 78.31%. This small difference in accuracy enhances the tool for assessing risk in crucial environments.

Furthermore, the presence of patterns suggests that, in the medium term, models with higher precision can be developed using this dataset and the latest image-based pattern recognition technologies. This demonstrates that ResNet18, with its convolutional layers, was more effective and practical than more complex models. Using

Grad-CAM, consistency was observed in the pixels determining the critical and normal classifications of an IP, confirming the model's focus on relevant image parts for classification decisions.

This study introduces a novel approach by focusing on the historical behavior of IP addresses rather than real-time traffic classification. In organizations with network access control, detecting potential behaviors and identifying scenarios where a device might be compromised by malware performing unexpected actions is crucial. By analyzing historical data, the system provides a more comprehensive view of user behavior over time, enhancing the ability to detect anomalies and potential threats.

The implications of these findings are significant: automated threat detection and reduced false positives enhance network management efficiency. The system's predictive intelligence optimizes decision-making by providing precise and detailed data-driven insights. Furthermore, operational costs are minimized by reducing losses from incidents and improving resource efficiency. This innovative approach not only strengthens network security but also serves as a robust tool for proactive threat management, rooted in behavioral sciences. The notable performance of the ResNet18 model, particularly when using Vortex Feature Positioning, underscores its practical potential in this field.

1.6 FUTURE WORKS

It is proposed that future work should focus on the following areas to enhance and expand the current study:

- **Implementation of models for converting tabular data into images**: It is suggested to use the DeepInsight [33], Refined [34], and IGTD [35] models to convert tabular data from this dataset into images. Comparing their performance against traditional tabular methods will provide valuable insights into their effectiveness.
- **Exploration of an end-to-end approach using convolutional networks**: It is recommended to investigate an end-to-end approach that employs convolutional networks and transposed convolutions to derive image representations. This exploration aims to determine whether this method yields superior results in image generation from tabular data.
- **Generation of new behavior models and database expansion**: It is proposed to use Generative Adversarial Networks (GANs) to generate new behavior models and expand the existing database [36].
- **Experiments with Kolmogorov-Arnold Networks for image classification**: It is suggested to conduct experiments using Kolmogorov-Arnold Networks for image classification [37].

DATA AVAILABILITY

The original dataset is publicly available at Ref. [4], while the generated dataset is available upon request.

CONFLICT OF INTEREST

The authors declare that they have no conflicts of interest.

ETHICS DECLARATIONS

This research was conducted with the highest ethical standards. The data were anonymized to ensure user privacy and no personally identifiable information (PII) was used. All procedures complied with relevant regulations and institutional guidelines.

ACKNOWLEDGMENTS

We extend our gratitude to Dr. Arash Habibi Lashkari of the Behaviour-Centric Cybersecurity Center (BCCC) at York University for sharing the dataset used in this project. We also thank Omar Arturo Diaz Alarcón Aguilar from the Universidad Popular Autónoma del Estado de Puebla (UPAEP) for his valuable contribution in providing preprocessing steps for the dataset.

REFERENCES

[1] Mellia, M., Carpani, A., & Cigno, R. (2002). Measuring IP and TCP behavior on edge nodes. Global Telecommunications Conference, 2002. *GLOBECOM '02. IEEE*, 2533–2537 vol.3 . https://doi.org/10.1109/GLOCOM.2002.1189087

[2] Callegari, C., Giordano, S., & Pagano, M. (2009). New statistical approaches for anomaly detection. *Secur. Commun. Netw.*, 2, 611–634. https://doi.org/10.1002/sec.104

[3] Shafi, M., Lashkari, A. H., Rodriguez, V., & Nevo, R. (2024). Toward generating a new cloud-based distributed denial of service (DDoS) dataset and cloud intrusion traffic characterization. *Information*, 15(4), 195. MDPI AG. https://doi.org/10.3390/info15040195

[4] BCCC-Dataset. BCCC CPacket Cloud-based DDoS. (2024). Behaviour-Centric Cybersecurity Center (BCCC). Available online: www.yorku.ca/research/bccc/ucs-technical/cybersecurity-datasets-cds

[5] Demaine, E., López-Ortiz, A., & Munro, J. I. (2002). Frequency estimation of internet packet streams with limited space, 348–360, https://doi.org/10.1007/3-540-45749-6_33.

[6] Ahmed, N., Ahmed, N., & Rajput, A. (2003). TCP/IP protocol stack analysis using MENet. *TENCON 2003. Conference on Convergent Technologies for Asia-Pacific Region*, 4, 1329–1333. https://doi.org/10.1109/TENCON.2003.1273132

[7] Kumar, R., & Kaur, T. (2014). Machine learning based traffic classification using low level features and statistical analysis. *Int. J. Comp. Appl.*, 108, 6–13. https://doi.org/10.5120/18961-0290

[8] Boger, M., Liu, T., Ratliff, J., Nick, W., Yuan, X., & Esterline, A. (2016). Network traffic classification for security analysis. *SoutheastCon*, 2016, 1–2. https://doi.org/10.1109/SECON.2016.7506668

[9] Chen, Z., Yan, Q., Han, H., Wang, S., Peng, L., Wang, L., & Yang, B. (2017). Machine learning based mobile malware detection using highly imbalanced network traffic. *Inf. Sci.*, 433–434, 346–364. https://doi.org/10.1016/j.ins.2017.04.044

[10] Stevanovic, M., & Pedersen, J. (2015). An analysis of network traffic classification for botnet detection. *2015 International Conference on Cyber Situational Awareness, Data Analytics and Assessment (CyberSA)*, 1–8. https://doi.org/10.1109/CYBE RSA.2015.7361120

[11] Savchenko, V., Matsko, O., Vorobiov, O., Kizyak, Y., Kriuchkova, L., & Tikhonov, Y. (2018). Network traffic forecasting based on the canonical expansion of a random process. *East.-Eur. J. Enterpr. Technol.*, 3, 33–41. https://doi.org/10.15587/1729-4061.2018.131471

[12] Granato, G., Martino, A., Baiocchi, A., & Rizzi, A. (2022). Graph-based multi-label classification for wifi network traffic analysis. *Appl. Sci.*. https://doi.org/10.3390/app122111303

[13] Millar, K., Cheng, A., Chew, H., & Lim, C. (2018). Deep learning for classifying malicious network traffic. 156–161. https://doi.org/10.1007/978-3-030-04503-6_15

[14] Chapaneri, R., & Shah, S. (2020). Multi-level Gaussian mixture modeling for detection of malicious network traffic. *J. Supercomp.* 77, 4618–4638. https://doi.org/10.1007/s11227-020-03447-z

[15] Hwang, R., Peng, M., Nguyen, V., & Chang, Y. (2019). An LSTM-based deep learning approach for classifying malicious traffic at the packet level. *Appl. Sci.* https://doi.org/10.3390/APP9163414

[16] Shi, Z., Luktarhan, N., Song, Y., & Yin, H. (2023). TSFN: A novel malicious traffic classification method using BERT and LSTM. *Entropy*, 25. https://doi.org/10.3390/e25050821

[17] Chapaneri, R., & Shah, S. (2019). Detection of malicious network traffic using convolutional neural networks. *2019 10th International Conference on Computing, Communication and Networking Technologies (ICCCNT)* (pp. 1–6). https://doi.org/10.1109/icccnt45670.2019.8944814

[18] Rong, C., Gou, G., Cui, M., Xiong, G., Li, Z., & Guo, L. (2020). MalFinder: An ensemble learning-based framework for malicious traffic detection. *2020 IEEE Symposium on Computers and Communications (ISCC)* (p. 7). https://doi.org/10.1109/ISCC50 000.2020.9219609

[19] Özdel, S., Ates, Ç., Ates, P. D., Koca, M., & Anarim, E. (2022). Payload-based network traffic analysis for application classification and intrusion detection. *2022 30th European Signal Processing Conference (EUSIPCO)* (pp. 639–642).

[20] Ren, G., Cheng, G., & Fu, N. (2023). Accurate encrypted malicious traffic identification via traffic interaction pattern using graph convolutional network. *Appl. Sci.* https://doi.org/10.3390/app13031483

[21] Cortes, C., & Vapnik, V. (1995). Support-vector networks. *Mach. Learn.*, 20(3), 273–297. https://doi.org/10.1007/BF00994018

[22] Breiman, L. (2001). Random forests. *Mach. Learn.*, 45(1), 5–32. https://doi.org/10.1023/A:1010933404324

[23] Quinlan, J. R. (1986). Induction of decision trees. *Mach. Learn.*, 1(1), 81–106. https://doi.org/10.1007/BF00116251

[24] Rish, I. (2001). An empirical study of the naive Bayes classifier. In *IJCAI 2001 Workshop on Empirical Methods in Artificial Intelligence*, 3(22), 41–46.

[25] Friedman, J. H. (2001). Greedy function approximation: A gradient boosting machine. *Ann. Stat.*, 29(5), 1189–1232. https://doi.org/10.1214/aos/1013203451

[26] Brodersen, K. H., Ong, C. S., Stephan, K. E., & Buhmann, J. M. (2010). The balanced accuracy and its posterior distribution. *2010 20th International Conference on Pattern Recognition* (pp. 3121–3124). https://doi.org/10.1109/ICPR.2010.764

[27] Chinchor, N. (1992). MUC-4 evaluation metrics. *Proceedings of the 4th Conference on Message Understanding* (pp. 22–29). https://doi.org/10.3115/1072064.1072067

[28] He, K., Zhang, X., Ren, S., & Sun, J. (2016a). Deep residual learning for image recognition. In *Proceedings of the IEEE conference on computer vision and pattern recognition* (pp. 770–778). https://doi.org/10.1109/CVPR.2016.90

[29] Howard, J., & Gugger, S. (2020). Fastai: A layered API for deep learning. *Information*, 11(2), 108. https://doi.org/10.3390/info11020108

[30] Park, J.-I., Seong, S., Lee, J., & Hong, C.-H. (2023). Vortex feature positioning: Bridging tabular IIoT data and image-based deep learning (Version 2). arXiv. https://doi.org/10.48550/ARXIV.2303.09068

[31] Benesty, J., Chen, J., Huang, Y., & Cohen, I. (2009). Pearson correlation coefficient. In *Noise reduction in speech processing* (pp. 1–4). Springer. https://doi.org/10.1007/978-3-642-00296-0_5

[32] Selvaraju, R. R., Cogswell, M., Das, A., Vedantam, R., Parikh, D., & Batra, D. (2017). Grad-CAM: Visual explanations from deep networks via gradient-based localization. In *Proceedings of the IEEE international conference on computer vision* (pp. 618–626). https://doi.org/10.1109/ICCV.2017.74

[33] Sharma, H., & Paliwal, K. K. (2020). DeepInsight: A methodology to transform a non-image data to an image for convolution neural network architecture. *Bioinformatics*, 36(20), 4815–4823. http://doi.org/10.1093/bioinformatics/btaa090

[34] Zheng, Y., Zhang, Q., Li, Z., Li, W., & Zhang, X. (2021). Refined: A refined image generation approach for tabular data using GANs. *IEEE Tran. Neur. Netw. Learn. Syst.*, 32(3), 1066–1076.

[35] Guo, J., Li, X., & Chen, D. (2021). IGTD: A novel method to convert tabular data into images for classification. *Patt. Recog.*, 111, 107701.

[36] Goodfellow, I., Pouget-Abadie, J., Mirza, M., Xu, B., Warde-Farley, D., Ozair, S., ... & Bengio, Y. (2014). Generative adversarial nets. In *Advances in neural information processing systems* (pp. 2672–2680). http://doi.org/10.1145/3422622

[37] Liu, Z., Wang, Y., Vaidya, S., Ruehle, F., Halverson, J., Soljačić, M., Hou, T. Y., & Tegmark, M. (2024). KAN: Kolmogorov-Arnold Networks (Version 4). arXiv. https://doi.org/10.48550/ARXIV.2404.19756

2 Cognition-Inspired Design in Mind

Demonstrations in Graphical Authentication

Jeremiah D. Still and Mary L. Still

2.1 OVERVIEW

We introduce cognition-inspired design (CID), an approach that prioritizes the understanding of both user and attacker cognitive in the development of cybersecurity systems. The approach is contextualized in the rich graphical authentication literature (Biddle, Chiasson, & Van Oorschot, 2012); in the present case, it is being used to aid the development of an alternative to traditional text-based passwords. Uniquely, within this approach, researchers focus on leveraging the power of human cognition. We begin by exploring the pressing issue of password overload, which has driven the need for more usable and secure authentication schemes. We then introduce graphical authentication as a potential solution to this challenge, highlighting its advantages and discussing various implementations. The unique security challenge of over-the-shoulder attacks (OSA), where an attacker can visually observe a user's password input, is highlighted throughout the chapter. An OSA scenario is particularly relevant as the cognitive capabilities of both the user and the attacker contribute to the effectiveness of this type of attack. We reflect on cognitive processes relevant to graphical authentication, namely human memory, attention, and object recognition. We discuss how these processes influence password selection, recall, and susceptibility to OSA.

DOI: 10.1201/9781003599142-3

To illustrate the practical application of CID, we present three demonstrations showcasing innovative graphical authentication designs. These examples show how applying cognitive understanding to the creation of authentication schemes can produce prototypes that are both usable and resistant to attacks. We highlight the need to consider the balance between usability and security when designing interfaces for cybersecurity systems such as next-gen authentication. This chapter offers readers an introductory understanding of the cognitive factors that influence graphical authentication and demonstrates the outcomes of the CID approach.

2.2 INTRODUCTION

The increasing number of online accounts and the corresponding demands for unique, complex passwords have placed a substantial cognitive burden on individuals, commonly experienced by end-users as password overload. Researchers conventionally reference the experience as password fatigue, password proliferation, or simply the problem of passwords. This password overload experience can result in insecure password practices (Cain, Edwards, & Still, 2018) such as the reuse of passwords across multiple accounts and the creation of weak, easily guessable passwords. These practices, in turn, expose users to significant security risks, including identity theft, financial fraud, and data breaches (Chiang & Chiasson, 2013).

Graphical authentication has emerged as a promising alternative to traditional text-based passwords, offering the potential to alleviate the cognitive strain associated with password memorability while maintaining, or even enhancing, security (Mator et al., 2020). By utilizing images, symbols, or patterns instead of alphanumeric characters, graphical authentication systems leverage the inherent strengths of human visual memory and recognition, making it easier for users to create and recall strong credentials (Still, Cain, & Schuster, 2017). Graphical authentication systems can be categorized into three main types: recognition-based, recall-based, and cued-recall (c.f., Cain & Still, 2018). Recognition-based systems present users with a set of images, requiring them to select the ones associated with their account (Hlywa, Biddle, & Patrick, 2011). Recall-based systems, on the other hand, prompt users to reproduce a previously seen image or sequence of images (Jermyn, et al., 1999). Cued-recall systems offer additional cues or hints to assist users in recalling the correct images (Biddle, Chiasson, & van Oorschot, 2008). The selection of a specific graphical authentication type depends on various factors, including the desired level of security, the target user population, and the context of use. Graphical authentication schemes have demonstrated effectiveness. For instance, Cain and Still (2018) found that 93% of participants remembered graphical passcodes after three weeks, while no participants remembered the traditional password over the same period. Findings like these suggest that graphical authentication systems can be particularly beneficial when users must remember their passwords for intermittent use.

However, graphical authentication is not without its challenges. One significant threat to the security of graphical authentication systems involves an attacker visually observing a user's authentication process to obtain their credentials illicitly (Cain, Werner, & Still, 2017). The visual nature of graphical passwords makes them particularly susceptible to OSA, as attackers can potentially discern passwords by simply watching users interact with the system. Addressing this challenge requires an understanding of both user and attacker cognition. One common design aspect to consider in developing these systems is mitigating the risk of visual observation while maintaining ease of use for the legitimate user.

A CID approach is essential to effectively addressing the challenges and opportunities associated with graphical authentication. This approach involves explicitly considering the cognitive processes of both users and attackers throughout the design process (e.g., prototyping, and redesign iterations). By understanding how humans perceive, process, and remember visual information, designers can create graphical authentication systems that are not only more usable and memorable for legitimate users but also more resistant to attacks like OSA.

2.3 COGNITIVE PROCESSES IN AUTHENTICATION

2.3.1 SHIFTING THE PARADIGM: THE END-USER AS A PARTNER IN SECURITY

Conventional wisdom in cybersecurity often portrays the end-user as the "weakest link" in the security chain, prone to errors and susceptible to manipulation. However, a growing body of research challenges this perspective, advocating for a more nuanced understanding of the user's role in security (Morris & Still, 2023; Zimmerman & Renaud, 2019). Rather than focusing solely on user vulnerabilities, it is crucial to recognize and leverage the cognitive strengths and the active role users play in maintaining effective security systems. Shifting to this perspective requires a deeper understanding of the cognitive factors that influence system usability and, ultimately, user compliance with existing best practices. To aid this transition, there is a need for designers to help fill the research gaps that exist between a theoretical understanding of cognitive capabilities and practical applications in this domain (e.g., Still, 2016). In taking a user-centered design approach to creating next-gen authentication, designers acknowledge that users are not passive recipients of security measures but active participants in their own security. By empowering users and leveraging their natural cognitive strengths, designers can create authentication systems that are both secure and usable.

2.3.2 MEMORY

Human memory plays a fundamental role in successful authentication experiences. Users are asked to remember their credentials to gain access to systems and services. However, human memory is fallible leading to forgotten credentials (Reason, 2000). It is clear that the complexity and sheer number of passwords users are expected to manage can easily become overwhelming (Choong & Greene, 2016). Cognitive overload can lead to insecure practices such as using the same password for different systems and creating weak passwords (Grawemeyer & Johnson, 2011).

Ironically, human long-term memory is hypothesized to be limitless, so it is not a "lack of space" that makes it difficult to remember credentials (c.f., Woods & Siponen, 2018). A variety of other factors impact successful recall and recognition.

At the most basic level, memory failures can occur because information was never adequately processed (encoding failure), it was not stored as a stable memory trace (consolidation failure), or it cannot be found (retrieval failure) (c.f., Miller, 2021). In the context of authentication, situational factors could thwart encoding and consolidation efforts. If feeling rushed or annoyed, users might spend less time and effort selecting a new credential. When asked to generate a new credential, credentials they already use will come to mind easily. Not only does this provide an obvious short-term solution to their "log-in problem" but users may feel that they will be more likely to remember a reused credential (compared to a novel credential; Tam et al., 2010; Woods & Siponen, 2024). If a user is distracted, attention may not be fully engaged in the encoding of the new credential. After creating a credential, users may continue to their next task (and then with the rest of their day), failing to devote time and effort to rehearsing the new credential. Even credentials that have been learned can be difficult to retrieve from memory. Effective retrieval depends not only on the quality of the memory trace, but also in having effective mechanisms for retrieval (Miller, 2021). High similarity between existing credentials and between applications requiring authentication can lead to interference (e.g., using a music service credential while trying to log in to another media service provider). Unique aspects associated with the credential itself or with the application requiring the credential may increase the odds of retrieval if that unique aspect can be used as a memory cue (e.g., Ensor et al., 2018). Even this benefit might be limited if that uniqueness was not attended to during encoding. In short, there are many reasons why a credential might be "forgotten".

Taking a CID approach, design of a successful authentication experience would consider demonstrated ways memory can fail and would work to mitigate those risks by identifying alternative designs that leverage known human strengths in memory. Graphical authentication systems provide a good demonstration of what can be accomplished with a CID approach. This type of authentication system leverages the *picture superiority effect*, a well-established cognitive phenomenon, to enhance password memorability (c.f., Adebimpe et al., 2023). The picture superiority effect refers to the finding that pictures are generally easier to remember than words (Paivio, 1979). Historically, this benefit has been attributed to dual coding, that is, the benefit for pictures emerges because both a visual representation of the picture as well as a verbal label for the picture are encoded. It is hypothesized that verbal representations are often spontaneously created for images (serving as a second memory code), but words seldom benefit from the spontaneous generation of a corresponding visual representation. From a mere statistical perspective, most models of long-term memory predict that having two memory traces associated with an item should improve the odds of retrieval compared to having only one memory trace. Alternative conceptualizations of the picture superiority effect suggest that images naturally have more distinctive characteristics than words; it is this distinctiveness that provides the comparative memory benefit for pictures (e.g., Ensor et al., 2018). In either case, by replacing text-based passwords with visual patterns, graphical authentication can tap into this memorability advantage, making it easier for users to create and retrieve complex credentials (e.g., strong passwords to resist brute force attacks).

2.3.3 ATTENTION

Attention has been described as a mechanism that selects information for further processing, filters out distracting or irrelevant information, and maintains information relevant for completing a search, task, or goal (Oberauer, 2019). This process is often characterized by its limitations – narrow focus and susceptibility to distraction – and by the way it is deployed – objects, space, location, and time. Attention plays a critical role in determining what information from the environment is brought into conscious awareness and, therefore, what information is available for further processing and even for later memory. In visual displays, humans have the ability to selectively pay attention to an object without being distracted by another object that appears in the same display (Johnston & Dark, 1986). This is particularly true if the distractor object is located farther away from the selected object. The ability to attend to objects in space can be extended to multiple objects. In the multiple object tracking (MOT) task participants focus on moving objects without being disrupted by similar distractors (Pylyshyn & Storm, 1988). Studies employing the MOT task demonstrate that participants can accurately track up to four targets simultaneously.

Attention can also be tuned to specific points in time. This *temporal attention*, is often tested using sequences or rapidly displayed items. Recognition of objects (and individual letters, numbers, and single words) can reliably take place within approximately 100 milliseconds. When presented in quick succession, briefly displayed items become subject to attentional limitations. For example, if an individual is searching for target items within a stream of briefly displayed items, they can suffer from an Attentional Blink (Potter, Chun, Banks, & Muckenhoupt, 1998) causing them to miss a target. Specifically, after identifying the first target in the stream, if a second target appears within 250 to 500 milliseconds, participants are likely to miss the second target altogether, as if attention had "blinked". It takes time for attention to disengage from processing the first target; if the second target appears during that time, attention is prevented from deploying. The attentional blink can be avoided by increasing the time between the targets and by inserting visual masks (stimuli consisting of lines or colors in nonmeaningful configurations) between the stimuli.

2.3.4 COGNITIVE WORKLOAD AND INTUITIVENESS

In an attempt to develop authentication systems with high levels of usability, designers consider how to make system use easier. One way to improve the objective and subjective ease of use is to decrease user cognitive workload. Cognitive workload is an index of the amount of resources needed to complete the cognitive processes associated with the primary task. These resources are finite and have capacity limitations; therefore, tasks requiring too many resources may produce errors. Cognitive workload is impacted by demands associated with the environment, situation, and task, as well as individual differences (see Longo et al., 2022, for a review). More specifically, cognitive workload tends to be higher in situations that involve divided attention, maintenance of multiple pieces of information in working memory, performance of mental transformations or comparisons, updating of mental models, and high levels of skill and knowledge, to name a few. Authentication mechanisms that impose a high cognitive load on users can lead to errors, frustration, and ultimately, a reluctance to use the system (see Kosch et al., 2023, for a review). In the context of graphical authentication systems, cognitive workload is particularly important to consider. Still and Cain (2020) found significant differences in the cognitive workload associated with different graphical authentication schemes. In this case, mental transformations required for the authentication process appeared to be the primary contributor to higher levels of cognitive load; entering a PIN or a traditional password would not impart this level of demand.

An interaction that can be completed effectively, fluently and with low cognitive effort is considered to be an intuitive interaction (e.g., Reinhardt & Huertienne, 2023). An intuitive authentication system is one that users can understand and interact with easily, without extensive instruction or training. Because of this, system intuitiveness is often a design goal. Intuitiveness has clear ties to cognitive workload, but it is also strongly impacted by memory. Repeated exposure to a particular interaction mapping (e.g., use the right button to increase the volume, not the left button; Still & Still, 2019) or a particular feature (e.g., find a picture and delete it, zoom in to see the details in a picture; Blackler et al., 2010) lead to the creation of a stable memory trace for the interaction. When the interaction is encountered again, it is likely to be completed more quickly and to engender a feeling of familiarity. When a novel design includes familiar features and interactions, the new design can benefit from those stored representations allowing for intuitive use as well (*intuitive first use*; Blackler et al., 2010). Accessing the memory for, and applying the appropriate response to, the familiar interaction can occur automatically and accurately and with little or no cognitive effort (e.g., Blackler et al., 2010; Still & Still, 2019). A feeling of fluency is generated during this process (Reinhardt & Huertienne, 2023).

To create an intuitive graphical authentication system, the interface should align with users' mental models and expectations. This can be achieved through the use of familiar images, symbols, and iconography; conventional and consistent interaction patterns; and familiar features. Designs that tap into these resources should be "intuitive". They have high learnability (if they require instruction at all) and should require little cognitive effort to use.

2.3.5 VISUAL PROCESSING

Visual salience, grouping principles, and object recognition are fundamental cognitive processes that play a crucial role in how users interact with graphical authentication systems. Visual salience refers to the degree to which an element stands out from its surroundings (Still, 2018). Uniqueness in color, contrast, or line orientation, for instance, can cause an object or a portion of a scene to be visually salient. Areas of high salience are thought to "call" for attention (c.f., Luck et al., 2020) thereby increasing the likelihood they will be fixated, encoded, and remembered. When designing interfaces, salience can be used to drive user attention quickly to a desired location (Still et al., 2020). In the context of graphical authentication systems, a visually salient passcode element might be detected more quickly – something the user might appreciate – but, this property also weakens security as an attacker's attention would also be drawn to the passcode element (Bulling, Alt, & Schmidt, 2012).

According to Bulling et al, 35% of password points were correlated with salient hotspot regions within a cued-recall scene used for authentication. If users tend to select password points (even unknowingly) based on salience, the systematicity associated with their password points creates a natural vulnerability. These findings provide a pointed demonstration of why cognitive capabilities must be considered from the perspective of the user and the attacker. Visual salience is a cue that would be available to both user and attacker, therefore, additional care must be taken to ensure critical information is protected.

Graphical authentication systems often use patterns, shapes, and objects as part of the authentication process. Visual design heuristics stemming from Gestalt principles (Arnheim, 1954) have long been used to predict how humans will interpret a visual interface. Gestalt grouping principles reflect our natural ability to make sense of the visual data that must be processed to successfully navigate our world. Some of the principles reflect a simplification of the visual world, while others almost reflect inferences about the visual world. The principles of *proximity*, *similarity*, and *closure*, for example, can be used to create visual patterns that are perceived as a cohesive unit rather than randomly co-occurring bits of data. The principle of *continuity* allows objects to be grouped along a line when a line is not visible; similarly, *reification* allows shapes to be perceived in the absence of closure and connected line segments. In graphical authentication, the convex hull graphical password scheme leverages these human perceptual capabilities, particularly closure and reification to create a low cognitive load interaction that is resistant to OSA (Anusha et al., 2018). The task requires users to find three passcode icons within a large display of decoys and then select inside the area of the invisible triangle formed by those three passcodes (Anusha et al., 2018; Sobrado & Birget, 2002). The user avoids revealing their passcode because they never overtly select the passcode icons.

The ability to recognize objects is supported by the same neurological architecture that allows for visual grouping in conjunction with visual long-term memory representations. Visual long-term memory is incredibly large, with a demonstrated capacity to store more than 2,500 detailed images (Brady et al., 2008)! Object recognition can occur quickly and with little effort. The process is robust to a variety of visual transformations that are commonly encountered in the environment, such as lighting changes, rotation, difference in depth, occlusion, and blurring. Because of this, for instance, a user could recognize a blurred object just like they could recognize a familiar object shrouded in fog in the distance. Similarly, a user can often recognize an object that is missing features just like they could find a pair of partially hidden shoes in a cluttered closet.

Taking a CID approach, it is important to consider the limitations of these processes. As a case in point, while it is true that objects can be recognized even if they are occluded or missing parts, cognitive research has revealed that some features are more important for object recognition than others. According to the Recognition-By-Components (RBC) theory, object features such as curvature and intersections, also known as nonaccidental properties, are needed for fast and accurate recognition of novel objects (Biederman, 1987). When those features are missing, an object may become unrecognizable. Interestingly, though, if an individual has had recent, unobstructed, exposure to the object they are able to correctly recognize an object even when the critical features have been deleted (Biederman, 1987). Previous exposure to an object, also makes it easier to recognize and identify the object if it is otherwise degraded upon encountering it again (Schacter, Chiu, & Ochsner, 1993; Snodgrass & Corwin, 1988). This priming effect plays a vital role in degraded object recognition. Priming, even for degraded objects, is experienced after a single exposure to a stimulus and has been shown to last 6 to 48 weeks (Cave, 1997; Snodgrass & Corwin, 1988).

2.4 THE DEVIL IS IN THE DETAILS

Designers have recognized that cognitive capabilities can be leveraged to improve the authentication experience. They have developed authentication schemes that use visually rich pictures instead of simple text and numbers. They have developed systems that rely on recognition instead of recall. While these changes have helped the user, in some cases, they would also help attackers. It is no coincidence that OSA prevention work has paralleled the development of new interaction schemes using graphical authentication (Cain, Werner, & Still, 2017). With this in mind, designers have also used knowledge of perceptual processing to make it more difficult for attackers to visually decipher passcodes during selection. The What You See is Where You Enter (WYSWYE; Khot et al., 2012) authentication process requires users to mentally perform several spatial manipulations before selecting their passcode. First, users find four passcode targets within a 5 × 5 grid. Then, they mentally delete the row and column that do not contain any of the passcode targets. After that, they mentally picture how the new 4 × 4 grid would map onto a smaller 4 × 4 blank grid. Finally, the user selects four locations on the blank grid that correspond to the locations of their passcode targets after completing those mental transformations. This technique is resistant to OSA but learning how to complete the task and successfully completing the mental transformations is challenging (Cain & Still, 2018). This process lacked the intuitiveness needed for widespread use, particularly for an interaction experience that is so important and pervasive as authentication.

2.5 COGNITION-INSPIRED DESIGN IN GRAPHICAL AUTHENTICATION

CID is an approach to system development that explicitly considers the cognitive capabilities of both users and attackers. By understanding how humans perceive, process, and remember information, designers can create systems that are more intuitive, usable, and secure. Still, Cain, and Schuster (2020) emphasize that an understanding of the human factors underlying authentication system interactions is crucial for developing systems that are secure and usable. To take full advantage of cognitive capabilities (strengths and potential limitations) several aspects of cognition must be considered, going beyond improving memorability. Factors such as memory, attention, workload, perception, and decision-making processes should be considered in designing for legitimate ease of use. It is just as important to consider how cognitive principles might be used to decrease an attacker's ability to guess or observe authentication credentials. The following demonstrations provide examples of CID applied to graphical authentication system development. The first two were inspired by research findings in object recognition, attention, and memory. The last demonstration was inspired by the attention literature, particularly findings related to multiple visual object tracking.

2.5.1 DEMONSTRATION 1: IMAGE FEATURE MANIPULATION

One example of CID in graphical authentication is the use of image feature manipulations to strike a balance between preventing OSAs and maintaining passcode recognition for the legitimate user (Hayashi, Dhamija, Christin, & Perrig, 2008). In OSA, an attacker attempts to observe a user's graphical password in a shared space by glancing at the interface during the authentication process. Ideally, a successful design would allow users to benefit from the use of images while simultaneously preventing attackers from using the images. A popular example of this is Use Your Illusion (UYI; Hayashi, et al., 2008). In this authentication scheme, users are assigned three pictures to serve as their passcode. During authentication, degraded versions of pictures are presented within a 3 × 3 grid. The pictures only retain their global colors and shapes, with no details. Users directly select their passcode pictures to authenticate. The degraded pictures are functionally meaningless to attackers, and because they are not easily identifiable, they are less likely to be accurately remembered. In contrast, users can easily recognize their passcode pictures because they generate a strong familiarity signal based on their memory for the originally studied, nondegraded version of the pictures.

In a similar effort, Tiller et al. (2019) investigated the use of image blurring as a technique to protect against OSA. With this approach it is possible to degrade an image too much, making it completely unusable. Therefore, the level of degradation must strike a balance between supporting legitimate users and preventing casual OSAs. For this reason, several levels of blurring were examined. Participants in the study were also tasked with playing the role of user and attacker. The goal of the study was to find the crossover point in the amount of distortion needed to prevent attackers from recognizing the image while also permitting legitimate use. Results revealed that moderate blurring (i.e., level 10) allowed all users to recognize the objects, but no attackers to recognize the objects. A slightly stronger blurring of the objects dropped legitimate user recognition to 72% accuracy (i.e., level 13). The latter finding demonstrates how even small value changes in image distortion can dramatically impact the performance of UYI. As expected from the memory literature, knowing what the undistorted image looks like has a significant impact on performance. A casual attacker unfamiliar with the original object was not able to determine the distorted object's identity (i.e., 90% of the time). However, users familiar with the objects were able to overcome a significant amount of distortion. Tiller et al.'s (2019) finding demonstrates the potential of CID to improve the security of graphical authentication systems. Simply manipulating visual features in images allows legitimate users to capitalize on their stored memory of the undistorted images but it can completely thwart attackers as they do not have those undistorted images in memory.

2.5.2 DEMONSTRATION 2: RAPID SERIAL VISUAL PRESENTATION

Another approach to employing CID in graphical authentication design is to use the rapid serial visual presentation (RSVP) scheme. RSVP demonstrates the potential of CID to enhance the security of graphical authentication by leveraging insights from attention and object recognition research. By itself, quickly displaying a sequence of items, one on top of another, creates an attention-demanding situation for observers. Cain and Still (2017) used the RSVP procedure in combination with feature degradation to make it difficult for an attacker to reconstruct the passcode. Uniquely, rather than using blurring for degradation, the line drawings used for authentication were degraded by removing key features that are critical for object recognition (e.g., nonaccidental properties; Biederman, 1987). In this graphical authentication system, users were assigned four target images (complete line drawings). During authentication, the degraded versions of the targets were presented in RSVP amongst other degraded line drawings. Users select their passcode targets by tapping the screen on small mobile devices. Like the UYI, this technique is based on recognition, not recall. The combined use of RSVP and objects that were missing critical features made this system highly resistant to OSAs. The results revealed that no OSA was able to capture an entire passcode; although half were able to recognize one of the four targets used in the passcode. As for legitimate users, all were able to login within the allotted three attempts. Overall, 84% of login attempts were successful. This example shows how understanding the nuanced interactions between attention, object recognition, and memory can lead to strategically created design features that make passcodes both secure and memorable.

2.5.3 DEMONSTRATION 3: CAMOUFLAGE

To resist OSAs, it is critical that users be able to hide their selection of passcode elements. As seen in the RSVP scheme, even briefly displayed, fragmented images can occasionally be detected in an OSA. In this light, CID was employed to develop an Incognito selection technique (Still & Bell, 2018). One way graphical authentication developers can hide a passcode is by camouflaging the passcode amongst distractors. Grid layouts of images are common in next-gen graphical authentication (e.g., UYI, Passfaces); they are also experienced conventionally during Personal Identification Number (PIN) based authentication. While the grid interaction used for PIN entry is ubiquitous and intuitive, it is amazingly suspectable to casual OSAs (e.g., ATM, physical entry gates). Still and Bell (2018) offered *Incognito* as a selection technique meant to camouflage passcode selection within grid layouts. In this instantiation, Incognito was assessed in the context of PIN entry on a screen displaying a 10-digit keypad; images were not used for the passcode. When using this system, participants controlled a mouse cursor which disappears over the keypad and transforms into a box surrounding an individual numeric key. Thus, the user experience was that of the mouse highlighting – by outlining the "key" – of the number the mouse was hovering over. To prevent an observer from knowing which number key was active, other nonactive keys would also have borders that would blink on and off. This served to camouflage mouse movement and selections. The legitimate user is able to map their own mouse movements with the changing borders in the grid, making it easy to know which box corresponds with the current mouse placement. The MOT literature shows that selectively attending to a single dynamic object within a complex scene is an easy task (c.f., Pylyshyn & Storm, 1988). This was evidenced by the data, as participants were able to log in using Incognito on 93% of the attempts. To test OSA resistance, causal attackers were provided optimal viewing of the interaction – they could see a video of the mouse movements along with a video of the screen. They were then permitted one guess of the PIN that was being selected. No participant playing the role of attacker was able to successfully determine the passcode. The selection technique leverages the user's attention toward their expected motor output. The attackers do not have access to the implicit motor output making it very difficult to discriminate between real and decoy selections.

2.6 BALANCING USABILITY AND SECURITY

In the design of graphical authentication systems, achieving a balance between usability and security is paramount. A system that is overly complex or difficult to use may deter users from adopting it or lead users to adopt strategies that compromise the security of the system. On the other hand, a system that prioritizes usability at the expense of security may leave users vulnerable to attacks. Researchers and designers have begun working to develop systems that balance usability and security and have met some success. Cain and Still (2018) conducted a study to evaluate the usability of various graphical authentication techniques designed to resist OSAs. Four primary prototypical OSA-resistant schemes were compared: grouping, translating to another location, disguising, and gaze-based input. All the schemes were found to be memorable and resistant to OSAs. Incredibly, after a three-week delay, memorability performance remained near the ceiling for all four schemes. Despite promising memory outcomes, learnability was identified as a weakness for these graphical schemes. Cognitive workload required to authenticate while using these schemes is another factor in need of further examination.

Unfortunately, general usability benchmarking beyond Cain and Still (2018) is limited. There are also limitations in the type of data that are reported in the current literature. This was highlighted by Mator et al. (2020) who note that next-gen authentication researchers typically only focus on efficiency and error rates leaving other key metrics untested. It is also uncommon to find reports that examine the performance of both legitimate users and attackers. OSA rates are reported on the security integrity side of the equation. Unfortunately, there is a lack of consistency in reporting and standardized procedures. The lack of benchmarking and limited measurement makes direct comparisons of the interaction schemes difficult; it makes interpretation of the underlying processes supporting the interactions challenging; and it leaves an incomplete record of the full user experience. Hopefully, future work will fill these gaps and facilitate next-gen authentication theory development.

Human-centered design in authentication has a unique constraint in that maintaining security is the top priority. Without security the intended design is useless. While usability is a critical consideration in the design of graphical authentication systems, it cannot come at the expense of security. To ensure the effectiveness of graphical authentication, it is essential to consider various threat models and attack vectors, such as shoulder surfing. In this chapter, we focused on preventing OSAs, which attempt to stop casual observers within a shared space. Leveraging human cognitive capabilities associated with memory for images, attention, and object recognition can mitigate the risk of OSA's while promoting legitimate uses. The field of cybersecurity emphasizes the importance of considering the threat model and proposed solution limitations. The CID approach naturally encompasses both of these aspects. By anticipating potential attack scenarios and understanding the limitations of graphical authentication techniques, designers can develop more secure systems for their unique user needs across various attack vectors.

2.7 CONCLUSION

As the digital landscape evolves, the need for secure and usable authentication becomes increasingly critical. CID offers a different perspective that considers the cognitive capabilities of users *and* attackers. These capabilities should be considered throughout the development process: during prototyping and when attempting to problem-solve possible solutions during development iterations. Graphical authentication has the potential to leverage naturally occurring human strengths to strike a balance between usability and security. A more advanced application of what is known about human memory, attention, cognitive workload, perception, and decision making are paramount in this endeavor. Interactions between cognitive processes are important to consider as well. For example, improving the memorability for the user with a new authentication scheme could result in unacceptable levels of cognitive load or, inadvertently result in a system that is easily compromised. The successful application of CID, from the perspective of promoting user success or preventing attackers, requires expertise in cognitive processes. Therefore, we suggest that computer scientists, engineers, and UX designers partner with cognitive scientists or human factors specialists to provide that knowledge base.

By adopting a CID approach, developers can create graphical authentication systems that are not only intuitive and easy to use but are also resistant to various attacks, including OSA attacks. The examples discussed in this chapter demonstrate the potential of CID to inform the design of novel and effective graphical authentication techniques. The CID approach will provide a useful framework as we attempt to create usable and secure design solutions. However, ongoing research and development are needed to address the evolving threats to authentication security and to explore the full potential of the proposed CID approach. Researchers also need to explicitly extend and examine the effectiveness of CID within other areas of cybersecurity.

ACKNOWLEDGMENT

The authors employed Gemini Advanced, an AI language model, to provide a first draft of the chapter's glossary and overview section. It was also used to check grammar and edit subsections by offering alternative statements after being directed to the authors' previous publications on related topics.

GLOSSARY

- **Authentication:** The process of verifying the identity of a user attempting to access a system or resource.
- **Biometrics:** The measurement and analysis of unique physical or behavioral characteristics (e.g., fingerprints, facial patterns, iris scans) for identification and authentication purposes.
- **Cognition:** The mental processes involved in acquiring knowledge and understanding, including thinking, remembering, perception, and problem-solving.
- **Cognitive Load:** The amount of mental effort required to perform a task.
- **Cognition-Inspired Design (CID):** A design approach that explicitly considers the cognitive capabilities and limitations of users and attackers to create more usable and secure systems.
- **Cued-Recall Graphical Authentication:** A type of graphical authentication where users are presented with cues or hints to help them recall the correct image or sequence of images.
- **Gestalt Principles:** A set of laws describing how humans perceive visual stimuli as organized patterns or whole forms, rather than individual elements.
- **Graphical Authentication:** A method of user authentication that uses images, symbols, or patterns instead of traditional text-based passwords.
- **Multifactor Authentication (MFA):** A security mechanism that requires users to provide multiple pieces of evidence to verify their identity, typically combining something they know (e.g., passcode), something they have (e.g., phone), and/or something they are (e.g., fingerprint).
- **Over-the-Shoulder Attack (OSA):** A type of attack where an attacker visually observes a user's authentication process to steal their credentials.
- **Password Overload:** The cognitive burden of managing numerous online accounts and passwords, leading to insecure practices and significant security risks.
- **Password Points:** A graphical authentication schemes term referring to specific locations or coordinates on an image that a user selects to compose their unique authentication credential.
- **Recall-Based Graphical Authentication:** A type of graphical authentication where users are required to reproduce a previously seen image or sequence of images from memory.
- **Recognition-Based Graphical Authentication:** A type of graphical authentication where users are presented with a set of images and required to select the correct one(s) associated with their account.
- **Security:** The protection of information and systems from unauthorized access, use, disclosure, disruption, modification, or destruction.
- **Usability:** The extent to which a product or system can be used by specified users to achieve specified goals with effectiveness, efficiency, and satisfaction in a specified context of use.
- **User-Centered Design:** A design approach that focuses on understanding and meeting the needs of users throughout the design and development process.

REFERENCES

Adebimpe, L. A., Ng, I. O., Idris, M. Y. I., Okmi, M., Ku, C. S., Ang, T. F., & Por, L. Y. (2023). Systemic literature review of recognition-based authentication method resistivity to shoulder-surfing attacks. *Applied Sciences, 13,* 10040.

Anusha, S., Shobha, C., Nannoji, Ranjini, P., Tejashwini, T. V., & Vishwesh, J. (2018). Graphical password with convex hull. *Proceedings of the 3rd National Conference on Image Processing, Computing, Communication, Networking, and Data Analytics,* 1–6.

Arnheim, R. (1954). *Art and visual perception: A psychology of the creative eye.* Berkeley, CA: University of California Press.

Biddle, R., Chiasson, S., & Van Oorschot, P. C. (2012). Graphical passwords: Learning from the first twelve years. *ACM Computing Surveys (CSUR), 44,* 19.

Biederman, I. (1987). Recognition-by-components: A theory of human image understanding. *Psychological Review, 94,* 115–147.

Blackler, A. L., Popovic, V., & Mahar, D. (2010). Investigating users' intuitive interaction with complex artefacts. *Applied Ergonomics, 41,* 72–92.

Brady, T. F., Konkle, T., Alvarez, G. A., Oliva, A. (2008). Visual long-term memory has a massive storage capacity for object details. *PNAS, 105,* 14325–14329.

Bulling, A., Alt, F., & Schmidt, A. (2012). Increasing the security of gaze-based cued-recall graphical passwords using saliency masks. *Proceedings of the SIGCHI Conference on Human Factors in Computing Systems,* 3011–3020.

Cain, A. A., Edwards, M. E., & Still, J. D. (2018). An exploratory study of cyber hygiene behaviors and knowledge. *Journal of Information Security and Applications, 42,* 36–45.

Cain, A. A., & Still, J. D. (2017). RSVP a temporal method for graphical authentication. *Journal of Information Privacy and Security, 13,* 226–237.

Cain, A. A., & Still, J. D. (2018). Usability comparison of over-the-shoulder attack resistant authentication schemes. *Journal of Usability Studies, 13,* 196–219.

Cain, A. A., Werner, S., & Still, J. D. (2017). Graphical Authentication Resistance to Over-the-Shoulder-Attacks. *In Proceedings of the 2017 CHI Conference Extended Abstracts on Human Factors in Computing Systems* (pp. 2416–2422).

Cave, C. B. (1997). Very long-lasting priming in picture naming. *Psychological Science, 8,* 322–325.

Chiang, H., & Chiasson, S. (2013). Improving user authentication on mobile devices: A touchscreen graphical password. *Proceedings of MobileHCI,* 1–10.

Chiasson, S., Forget, A., Biddle, R., & van Oorschot, P. C. (2008). Influencing users towards better passwords: Persuasive Cued Click-Points. *In British HCI Annual Conference,* 121–130.

Choong, Y., & Greene, K. K. (2016). What's a special character anyway? Effects of ambiguous terminology in password rules. *Proceedings of the Human Factors and Ergonomics Society Annual Meeting, 60,* 760–764.

Ensor, T. M., Surprenant, A. M., & Neath, I. (2018). Increasing word distinctiveness eliminates the picture superiority effect in recognition: Evidence for the physical-distinctiveness account. *Memory & Cognition, 47,* 182–193.

Grawemeyer, B., & Johnson, H. (2011). Using and managing multiple passwords: A week to a view. *Interacting with Computers, 23,* 256–267.

Hayashi, E., Dhamija, R., Christin, N., & Perrig, A. (2008). Use your illusion: Secure authentication usable anywhere. *Proceedings of the 4th symposium on Usable privacy and security,* 35–45.

Hlywa, M., Biddle, R., & Patrick, A. S. (2011). Facing the facts about image type in recognition-based graphical passwords. *In Annual Computer Security Applications Conference,* 149–158.

Jermyn, I., Mayer, A., Monrose, F., Reiter, M. K., & Rubin, A. (1999). The design and analysis of graphical passwords. *In USENIX Security Symposium*, 1–15.

Johnston, W. A., & Dark, V. J. (1986). Selective attention. *Annual Review of Psychology, 37*, 43–75.

Khot, R. A., Kumaraguru, P., & Srinathan, K. (2012). WYSWYE: Shoulder surfing defense for recognition based graphical passwords. *In Proceedings of the 24th Australian Computer Human Interaction Conference*, 285–294.

Kosch, T., Karolus, J., Zagermann, J., Reiterer, H., Schmidt, A., & Woźniak, P. W. (2023). A survey on measuring cognitive workload in human-computer interaction. *ACM Computing Surveys, 53*, 283.

Longo, L., Wickens, C. D., Hancock, G., & Hancock, P. A. (2022). Human mental workload: A survey and a novel inclusive definition. *Frontiers in Psychology, 13*, 883321.

Luck, S. J., Gaspelin, N., Folk, C. L., Remington, R. W., & Theeuwes, J. (2020). Progress toward resolving the attentional capture debate. *Visual Cognition*, doi: 10.1080/13506285.2020.1848949

Mator, J. D., Lehman, W. E., McManus, W., Powers, S., Tiller, L., Unverricht, J. R., & Still, J. D. (2020). Usability: Adoption, measurement, value. *Journal of Human Factors, 63*, 956–973.

Miller, R. (2021). Failures of memory and the fate of forgotten memories. *Neurobiology of Learning and Memory, 181*, 107426.

Morris, T.W., & Still, J.D. (2023). Cybersecurity hygiene: Blending home and work computing. In W. Patterson (Ed.), *New Perspectives in Behavioral Cybersecurity*. Boca Raton, FL: CRC Press.

Oberauer, K. (2019). Working memory and attention – a conceptual analysis and review. *Journal of Cognition, 2*, 36.

Paivio, A. (1979). *Imagery and Verbal Processes*. London, Ontario: Psychology Press.

Potter, M. C., Chun, M. M., Banks, B. S., & Muckenhoupt, M. (1998). Two attentional deficits in serial target search: The visual attentional blink and an amodal task-switch deficit. *Journal of Experimental Psychology: Learning, Memory, and Cognition, 24*, 979–992.

Pylyshyn, Z. W., & Storm, R. W. (1988). Tracking multiple independent targets: Evidence for a parallel tracking mechanism. *Spatial Vision, 3*, 179–197.

Reason, J. (2000). *Human error: models and management*. Cambridge University Press.

Reinhardt, D., & Hurtienne, J. (2023). Measuring intuitive use: Theoretical foundations. *International Journal of Human-Computer Interaction, 40*, 2453–2483.

Schacter, D. L., Chiu, C. Y. P., & Ochsner, K. N. (1993). Implicit memory: A selective review. *Annual Review of Neuroscience, 16*, 159–182.

Snodgrass, J. G., & Corwin, J. (1988). Perceptual identification thresholds for 150 fragmented pictures from the Snodgrass and Vanderwart picture set. *Perceptual and Motor Skills, 67*, 3–36.

Sobrado, L., & Birget, J-C. (2002). Graphical passwords. *The Rutgers Scholar, 4*. Retrieved from https://rutgersscholar.libraries.rutgers.edu/index.php/scholar/article/view/60

Still, J. D. (2016). Cybersecurity needs you! *Interactions, 23*, 54–58.

Still, J. D. (2018). Web page visual hierarchy: examining Faraday's guidelines for entry points. *Journal Computers Human Behavior, 84*, 352–359.

Still, J. D., & Bell, J. (2018). Incognito: Shoulder-Surfing resistant selection method. *Journal of Information Security and Applications, 40*, 1–8.

Still, J. D., & Cain, A. A. (2020). Over-the-Shoulder attack resistant graphical authentication schemes impact on working memory. In T. Ahram & E. Karwowski (eds) *Advances in Human Factors in Cybersecurity. AHFE 2019. Advances in Intelligent Systems and Computing, 960*, 79–86. Springer, Cham.

Still, J. D., Cain, A., & Schuster, D. (2017). Human-centered authentication guidelines. *Information & Computer Security, 25*, 437–453.

Still, J. D., Hicks, J. M., & Cain, A. A. (2020). Examining the influence of saliency in mobile interface displays. *AIS Transactions on Human-Computer Interaction, 12*(1), 28–44.

Still, M. L., & Still, J. D. (2019). Cognitively describing intuitive interactions. In A. Blackler (Ed), *Intuitive interaction: research and applications.* (pp. 41–61). Boca Raton, Florida, USA: Taylor and Francis.

Tam, L., Glassman, M., & Vandenwauver, M. (2010). The psychology of password management: A tradeoff between security and convenience. *Behaviour and Information Technology, 29*, 233–244.

Tiller, L., Cain, A., Potter, L., & Still, J. D. (2019). Graphical authentication schemes: Balancing amount of image distortion. In Ahram T., Nicholson D. (Eds), *Advances in human factors in cybersecurity* (pp. 88–98). Orlando, FL: Springer.

Woods, N., & Siponen, M. (2018). Too many passwords? How understanding our memory can increase password memorability. *International Journal of Human-Computer Studies, 111*, 36–48.

Woods, N., & Siponen, M. (2024). How memory anxiety can influence password security behavior. *Computers & Society, 137*, 103589.

Zimmermann, V., & Renaud, K. (2019). Moving from a 'human-as-problem" to a 'human-as-solution" cybersecurity mindset. *International Journal of Human-Computer Studies, 131*, 169–187.

3 Cybersecurity Defense
Against Pro or Amateur Attackers

Wayne Patterson

The content of this chapter was originally given as an invited seminar at one of the largest senior citizen communities in a nearby Maryland neighborhood of Washington, DC.

3.1 DEFINITIONS

This is not the usual practice in the cybersecurity world, but for a long time now, I have spoken and written about the challenges in cybersecurity as being defined as a contest between "Attackers" and "Defenders".

And in trying to classify both, I find it useful to define two groups:

Professionals – highly skilled, substantial if not unlimited resources, and
Amateurs – lesser skills, few or no resources

3.2 THINK OF THE CHALLENGE AS "GAME THEORY"

You have opponents who are playing a game according to certain "rules" (stated or implicit). On one side you have the *attackers*, trying to gain something from you; and on the other side, *defenders* (like you), trying to protect your (cyber) environment; and on both sides, you have very skilled and well-resourced players (*the pros*) and also the less skilled and with fewer resources (*the amateurs*).

 DOI: 10.1201/9781003599142-4

3.3   EXAMPLE: COLONEL BLOTTO

One of the sources for examples in game theory arises from military strategy. One classic military strategy game is called Colonel Blotto [1].

Colonel Blotto must defend two mountain passes, and to do this, she has three divisions. She can successfully defend one of her mountain passes if she has a division that is pitted against an enemy unit of equal or smaller strength. The enemy, however, has only two divisions. Blotto will lose the battle if either pass is captured. Neither side has any advance information on how the opponent's divisions are allocated. The game, then, is described by the different possible alignments of both Blotto's divisions and the opponents.

The strategies for Blotto have four possibilities, which we describe as (3,0), (2,1), (1,2), and (0,3). The first coordinate describes how many divisions Blotto assigns to the first mountain pass, and the second coordinate the number of divisions assigned to the second pass. Clearly it wouldn't make any sense for Blotto to leave any of her divisions behind, so in each case Blotto is assigning all three divisions. The opponent, as indicated above, has only two divisions and thus has only three potential strategies: (2,0), (1,1), and (0,2).

Develop the game matrix: Let's also assume that the payoff for the game is one unit depending on who wins. As an example, if Blotto assigns all of her divisions to the first pass, that is (3,0); and the opponent assigns one of his divisions to each pass, otherwise (1,1), the colonel loses the battle and the opponent wins, thus the matrix entry is –1.

The battle is lost if either pass is captured. The payoff matrix is thus:

		Enemy		
		(2,0)	(1,1)	(0,2)
	(3,0)	1	−1	−1
Blotto	(2,1)	1	1	−1
	(1,2)	−1	1	1
	(0,3)	−1	−1	1

A row or column may also be removed if it is dominated by a probability combination of other rows or columns.

Just one preliminary observation of Colonel Blotto's game, her (3,0) and (0,3) strategies are never required, since the enemy only has two divisions, and Blotto can successfully defend any pass with only two divisions.

But we can reach the same conclusion through examination of the payoff matrix. Because the payoffs in the (3,0) row are all lower or equal to the numbers in the same column in the second row, and similarly for the (0,3) row; by the principle of domination, we can reduce the game to:

Colonel Blotto game "reduced"

		Enemy		
		(2,0)	(1,1)	(0,2)
Blotto	(2,1)	1	1	−1
	(1,2)	−1	1	1

A row or column may also be removed if it is dominated by a probability combination of other rows or columns.

Now, examining the game matrix from the colony perspective, that is the Enemy's perspective, the strategy (1,1) is also dominated. Thus we can also remove that column from the game.

Colonel Blotto game second reduction:

		Enemy	
		(2,0)	(0,2)
Blotto	(2,1)	1	−1
	(1,2)	−1	1

Considering the remaining 2 × 2 matrix, we see we have essentially reduced the Colonel Blotto game to the game of matching coin flips, and from our previous analysis we realize that the players should each randomize their choice of strategy leading to a 50–50 conclusion.

3.4   EXAMPLE: GAME MODEL

	Pro	Amateur
Pro	CIA, NSA, KGB, Mossad, IRA[a], WikiLeaks, Stuxnet#	Hacker in third-world developing country, Robert Morris (?)[b]
Amateur	All of us tomorrow?	All of us here today

a   IRA = Internet Research Agency **NOT** the Irish Republican Army.
b   See next slides.

3.5   STUXNET

The year 2010 produced a "game changer". For perhaps the first time, a malicious hardware and software attack coined as Stuxnet [2], infected nuclear facilities in Iran.

One critical difference here was that *previous malware always* was produced by individuals or small groups, sought random targets, and caused relatively minimal damage.

Stuxnet was discovered by security researchers. It was determined to be a highly sophisticated worm that spread via Windows, and targeted Siemens software and equipment (it was known that Siemens SCADA was the software managing the Iranian nuclear power development systems).

What does Stuxnet do? Initially it just tries to access all computers possible (like previous worms). But then if the attacked computer is not a Siemens SCADA system, then no harm done.

If it is a SCADA system, then it infects the logic and could destroy the system

So what happened? Different versions of Stuxnet infected five Iranian organizations, presumably related to the uranium enrichment infrastructure. The

Symantec Corporation reported in August 2010 that almost 60% of Stuxnet infected computers worldwide were in Iran.

3.6   MORRIS WORM

On November 2 of that year, Robert Morris, then a graduate student in computer science at Cornell, created the worm in question and launched it on the Internet.

The worm could move from one machine to the other, and it was estimated that it infected eventually about 2000 computers within 15 hours [3].

The fix for each machine consisted only of deleting the many copies of the worm. The US Government Accountability Office estimated the cost of the damage in the range of $100,000–$10 million – obviously not a very accurate assessment.

Robert Morris was tried and convicted under the Computer Fraud and Abuse act, and was sentenced to three years' probation, 400 hours of community service, and fined $10,050. He was subsequently hired as a professor of computer science at the Massachusetts Institute of Technology, where he continues to teach and research to this day.

His father, Robert Morris Sr., had been a researcher for the National Security Agency, and because of his son's adventure, senior was obliged to resign from the NSA – what you might call the Cybersecurity Oedipus Complex.

3.7   GAME MODEL PAYOFF

	Pro	Amateur
Pro	$1000000, –$1000000	$0, –$100
Amateur	$100, 0	$100, –$100

3.8   TYPES OF PRO ATTACKERS

1. Ransomware
2. Phishing
3. Telephone scams
4. Global concerns, and so on

3.9   RANSOMWARE

A new type of attack, ransomware [4], made its first known appearance in 1989. These are attacks designed to disable a person's computer and information until a sum of money is paid to the perpetrator.

The first known case of ransomware is about as bizarre a story as in the entire annals of cybersecurity. It dates back to 1989 (pre-Web) and it was launched– on floppy disks – by an evolutionary biologist and Harvard PhD, Joseph Popp. His ransomware encrypted a victim's files after a certain number of reboots of their system.

He had mailed this virus to approximately 20,000 recipients on floppies.

3.10 CRYPTOLOCKER AND PETYA

Ransomware resurfaced in 2013 with CryptoLocker, which used Bitcoin to collect ransom money. It is estimated that the operators of Crypto Locker had procured about $27 million from infected users.

Another widespread version of ransomware is called Petya, originating in Russia. It also had a cloned version called Wannacry.

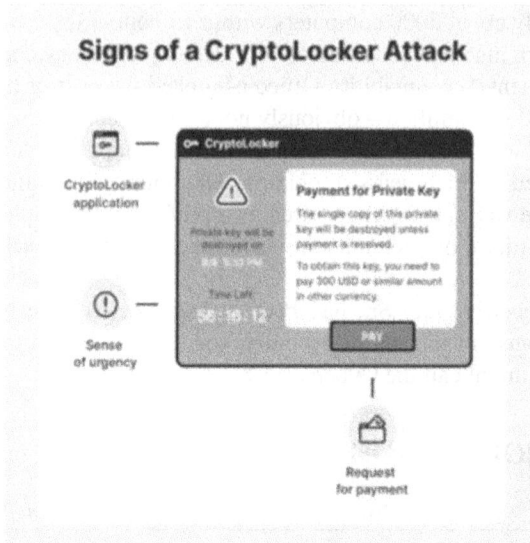

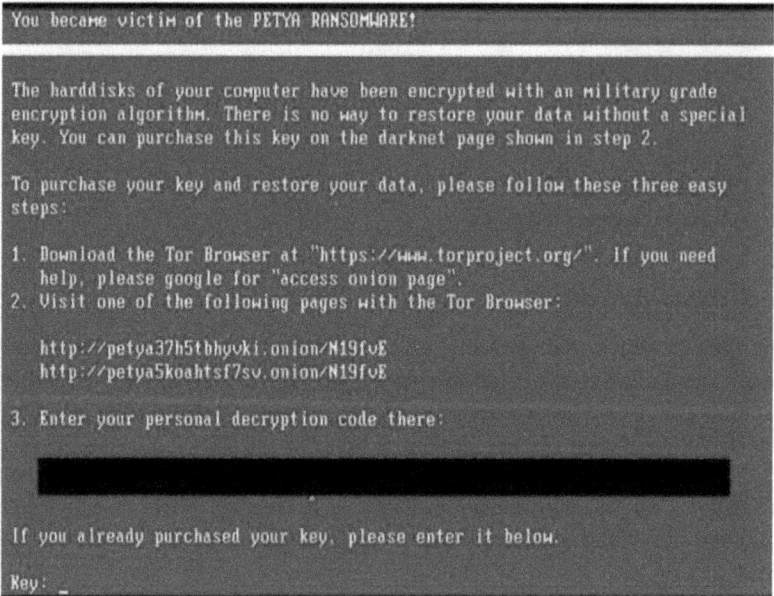

Signs of a CryptoLocker Attack

CryptoLocker application

Sense of urgency

CryptoLocker

Payment for Private Key

The single copy of this private key will be destroyed unless payment is received.

To obtain this key, you need to pay 300 USD or similar amount in other currency.

Time Left
56:16:12

PAY

Request for payment

```
You became victim of the PETYA RANSOMWARE!

The harddisks of your computer have been encrypted with an military grade
encryption algorithm. There is no way to restore your data without a special
key. You can purchase this key on the darknet page shown in step 2.

To purchase your key and restore your data, please follow these three easy
steps:

1. Download the Tor Browser at "https://www.torproject.org/". If you need
   help, please google for "access onion page".
2. Visit one of the following pages with the Tor Browser:

   http://petya37h5tbhyvki.onion/N19fvE
   http://petya5koahtsf7sv.onion/N19fvE

3. Enter your personal decryption code there:



If you already purchased your key, please enter it below.

Key: _
```

3.11 PHISHING

Another type of malware, "phishing", attempts to deceive a user into giving up sensitive information, most often by merely opening a "phishing" email, or even by responding to such an email by returning sensitive information;

Phishing attacks are a form of illicit software designed primarily to obtain information to benefit the "Phisher".

These attacks might occur from opening an email supposedly from some trusted source. The purpose for the attack might be an urgent request for money, and cause recipient to open an attachment.

Many users might not realize that opening a Word, Excel, PowerPoint document may contain within its code (called a macro) that may then infect her system.

Another approach in the phishing attack might be to encourage the recipient to follow a link which purports to require the user to enter an account name, password, or other personal information which then is being transmitted to the creator of the phishing attack.

One "phishing" example from 2016 involves an email allegedly from the PayPal Corporation. What follows is a screenshot of the attack itself. In this case, the objective of the attack is to fool the recipient into believing that this is a legitimate email from PayPal, asking for "help resolving an issue with your PayPal account,?" And consequently passing on the login information to the phishing attacker.

A prudent user would look carefully at the email address, service.epaypal@outlook.com and realize that it was a bogus email address.

On a larger scale, the use of malicious software and sometimes hardware in order to damage or destroy critical infrastructure.

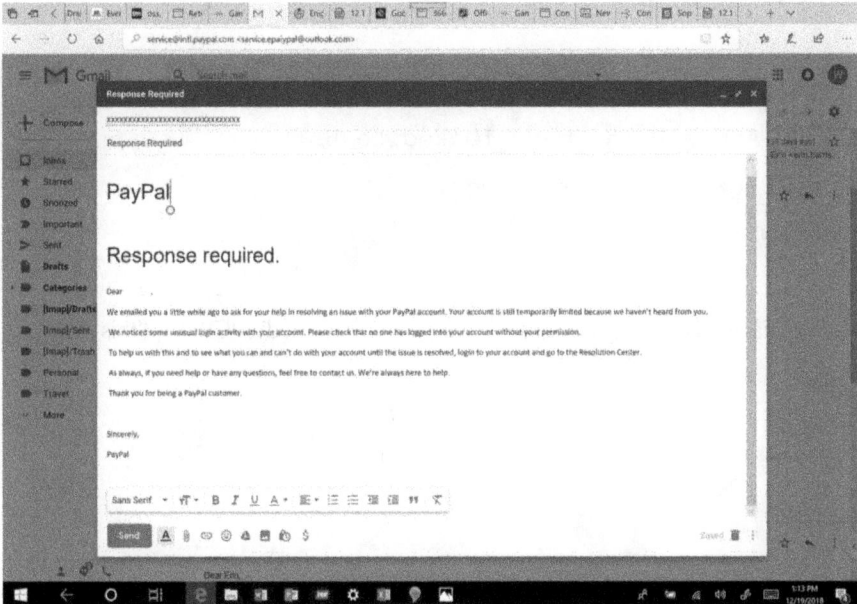

3.12   TELEPHONE SCAMS

Here are several examples one can observe in terms of telephone links that are potentially meant to lure the user into destructive links. These all appear in link messages and we will explain the clue to detecting them.

```
Fw: _psotfk17… Your package is scheduled for delivery,
#N 84672

.

.

Temu_Confirmatiion <68zu3bsstxkncnjnppng>
```

The clue here is that the legitimate person has an email address starting with "psotf" but with no "k17".

```
Save ap-ple2023eaccnt-8213912009

Saving the number will add a new contact
```

In this case, the use of "apple" is meant to suggest a legitimate company, but the give away is using "ap-ple".

```
AppleID#694717

Your AppleID has been_IOcked,to Unlock

it,Cl!ck this_link //_9qr.de/

HWt1DB
```

Here we have misspellings and bogus characters: "IOcked", "Cl!ck", "this_link".

← 👤 ups504w@packageinczpzka... ⋮

Sunday, Aug 20 • 2:18 PM

Subject: TIPFQVGQMUHWOTK

Your Parcel Has Been Updated

We've inform you about your parcel

Your parcel is on hold at UPS warehouse

Redelivery schedule required, Update your delivery address by click the link below:

\# reschedule-delivery.watest.us

If in next 24 hours your address not updated, your parcel will returned.

Kind Regard, UPS Team

2:18 PM

⊕ 📇   Text message   ☺ 🎤

The header in the top line suggests the sender is not really UPS: ups504w@ packageincz...; also the grammar "We've inform you".

USPS is the United States Postal Service. But in this case, the telephone area code, "+44" indicates the source is in the country code +44, in other words, the United Kingdom.

3.13　SOME GLOBAL EXAMPLES

3.13.1　ESTONIA AND WORLD WAR III

Estonia [5] is one of the three small so-called "Baltic States" in Northeastern Europe, on the Baltic Sea: Latvia, Lithuania, and Estonia. All were taken over by Russia when it formed the Soviet Union, or USSR.

On April 27, 2009, a series of cyberattacks occurred in the Estonian parliament, banks, government ministries, and media; so, the President of Estonia, Toomas Ilves, initially concluded that these attacks were coming from Russia, and so Estonia seriously contemplated declaring war on Russia.

This would have seemed foolish considering the relative sizes – Estonia about 1.3 million people versus Russia's 143.4 million – except that Estonia is and was a NATO member country, and the NATO constitution requires that if a NATO member declares war, all other NATO members are obliged to join in that declaration of war. (This is one, if not the main reason, that the Ukraine has been trying get into NATO.)

After consideration, Estonia concluded that there was insufficient evidence to blame the cyberattacks on the Russian government, so no state of war was declared.

Incidentally, (now former) President Ilves holds degrees from Columbia and Penn, and also taught at Simon Fraser University in British Columbia, Canada.

3.13.2   THE SONY PICTURES HACK: GUARDIANS OF PEACE (GOP)

SONY Pictures had decided to produce a film ridiculing North Korea's Kim Jong-Il. As it was nearing release, around November 24, 2014, an email was sent to SONY Pictures and many others. It contained the message that follows:

We will clearly show it to you at the very time and places The Interview be shown, including the premiere, how bitter fate those who seek fun in terror should be doomed to. Soon all the world will see what an awful movie Sony Pictures Entertainment has made. The world will be full of fear. Remember the 11th of September 2001. We recommend you to keep yourself distant from the places at that time. (If your house is nearby, you'd better leave.)

Also, terrorist actions were threatened at the premiere in NYC.

3.14   SUMMARY

As I hope this presentation has shown, for virtually anyone the issues regarding cybersecurity knowledge and awareness are important topics, not just for the "pros" and "amateurs" as have been described in this talk.

Even though we ordinary users are always more vulnerable to the most capable and well-resourced "pros", most of us are not big enough "fish" or threats for us to be concerned about such high-level attacks. Thus it is more important for most of us to gain enough experience and expertise to be able to fend off the low-level types of attacks – the "amateurs" in this terminology.

It is hoped that some of the tools described here can be useful to all in protecting yourselves – not against the pros – but against the amateurs whom you are more likely to encounter.

Thanks for listening,

Wayne Patterson

REFERENCES

[1] Blackett, D W (1958), Pure Strategy Solutions to Blotto Games. *Nav Res Log Quart* 5:107–109.

[2] Patterson, W and C Winston (2019), *Behavioral Cybersecurity*, CRC Press, pp. 31–33.

[3] Dressler, J (2007), "USA vs. Morris", *Cases and Materials on Criminal Law*. St. Paul, MN: Thomson/West. ISBN 978-0-314-17719-3.

[4] Patterson, W and C Winston (2019), *Behavioral Cybersecurity*, CRC Press, pp. 30–31.

[5] Wikipedia (2024), Estonia, en.m.wikipedia.org

Section II

Perspectives on Business
Approaches in Cybersecurity

4 Hybrid Intelligence in Cybersecurity Banking

Tihomir Dovramadjiev, Rusko Filchev, and Rozalina Dimova

4.1 INTRODUCTION

In modern society, the integration of hybrid intelligence (HI) in cybersecurity banking has become more and more important to protect the digital assets [1-6]. This chapter explores the psychological and technical aspects of HI in banking cybersecurity, emphasizing the need for both human factors (HFs) and advanced technologies to create a resilient defense system [7-15]. The dual approach of integrating psychological and technical perspectives, coupled with HI, is essential for creating a robust and sustainable cybersecurity framework in banking. By leveraging the strengths of both artificial intelligence (AI) and human intelligence, banks can effectively defend against cyber threats, ensuring the protection of digital assets and maintaining trust in financial systems. The interplay between psychological and technical aspects of HI in banking cybersecurity, highlighting the indispensable role of HF in correlation advanced technologies (Figure 4.1). The dual approach, which merges psychological experience with cutting-edge technology, is vital for developing a resilient and adaptable defense system. HFs such as awareness, vigilance, and the ability to respond to complex social engineering attacks are integral to this HI framework. Banks must harness these human elements, leveraging them alongside AI to build a robust cybersecurity infrastructure [16-19]. The principles of Society 5.0 and Industry 5.0 further underscore the importance of this integration [20-26]. Society 5.0 aims to create a super-smart society where technology and human ingenuity work together to

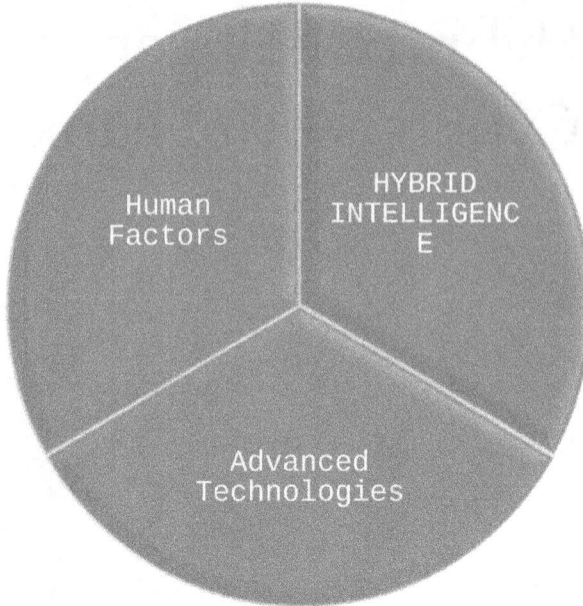

FIGURE 4.1   Hybrid intelligence in cybersecurity Banking, the invisible system of correlation of human factors and advanced technologies.

solve social problems, including cybersecurity challenges [27-30]. Similarly, Industry 5.0 emphasizes the synergy between human-centric approaches and advanced AI technologies to achieve sustainable and effective solutions. By incorporating the strengths of both human intelligence and AI, banks can enhance their ability to anticipate, detect, and counteract cyber threats, ensuring the protection of digital assets and upholding trust in financial systems.

Figure 4.1 represents the intricate and interwoven network where HFs—such as cognitive awareness, decision-making, and behavioral vigilance—interact with advanced technological tools like AI, machine learning, and sophisticated encryption methods. While the interactions between these elements may not be immediately visible, their combined impact is substantial and crucial for modern cybersecurity efforts. The "invisible system" emphasizes that despite its lack of physical presence, this integration of human and technological components forms the backbone of a resilient and adaptive cybersecurity framework. Understanding this invisible network is essential because it reveals how human insight and technological advancements collaboratively fortify defenses against cyber threats. This unseen synergy ensures that cybersecurity measures are comprehensive, dynamic, and capable of evolving with emerging threats. Figure 4.1 underscores that while the system's workings may be hidden from direct view, its real-world effectiveness and importance are profound, demonstrating that a sophisticated, invisible yet real framework is indispensable for safeguarding digital assets in today's interconnected and high-risk environment.

4.2 PSYCHOLOGICAL PERSPECTIVE ON CYBERSECURITY

Attackers and defenders in the realm of cybersecurity banking operate under distinct psychological motivations and responsibilities [31-35]. Attackers, driven by various motivations, engage in cybercrime at different levels. Attackers in the realm of banking cybersecurity are motivated by diverse factors, each operating on different scales with unique objectives and techniques. On an individual level, these cybercriminals are often driven by the desire for monetary gain, the need for recognition among their peers, or the intent to exact revenge for personal grievances. They use a variety of methods to siphon funds, achieve notoriety, or disrupt systems to avenge perceived wrongs. On a broader scale, national-level attackers engage in cyber espionage or attempts to destabilize the financial infrastructure of competing nations. Such attacks, often state-sponsored or politically motivated, aim to collect sensitive information, interfere with financial operations, or erode the economic confidence of the targeted countries. On a more localized level, regional cybercrime organizations concentrate their efforts on regional banks and financial systems, exploiting their deep understanding of local institutions and regulatory environments to identify and exploit weaknesses. These groups typically operate within specific geographical areas, tailoring their strategies to effectively breach regional financial entities. Recognizing these varied motivations is essential for crafting effective defensive measures and addressing the wide range of threats posed by attackers in the cybersecurity domain.

On the other side, defenders, including banks and regular users, face significant responsibilities and challenges. Banks and financial systems are tasked with protecting customer data, maintaining trust, and ensuring compliance with stringent regulations. However, they face the ongoing challenge of keeping up with the rapidly evolving threat landscape and managing often limited cybersecurity budgets. Defenders in the cybersecurity, including both banks and financial systems, bear significant responsibilities and face numerous challenges. Banks and financial institutions are primarily tasked with protecting customer data, ensuring the integrity and confidentiality of sensitive information, and maintaining the trust of their clients. This responsibility extends to ensuring compliance with a myriad of regulations designed to safeguard financial transactions and personal data. However, these institutions face substantial challenges in fulfilling these duties. One of the foremost challenges is keeping pace with the rapidly evolving threat zone, where cyber threats are constantly becoming more sophisticated and harder to detect. Additionally, managing cybersecurity budgets effectively poses a persistent challenge. Allocating sufficient resources to cybersecurity measures without compromising other critical areas of the business is a delicate balancing act. This requires continual investment in advanced technologies, skilled personnel, and ongoing training to stay ahead of potential threats. In this dynamic and high-stakes environment, banks and financial systems must navigate these challenges to protect their customers and maintain the overall security and trust in the financial ecosystem. Regular users, meanwhile, have the responsibility to safeguard their personal information and engage in safe online practices. Unfortunately, many users struggle with a lack of awareness and are highly susceptible to social engineering attacks, making them vulnerable targets in the cybersecurity zone. This dynamic between attackers and defenders highlights the critical importance of

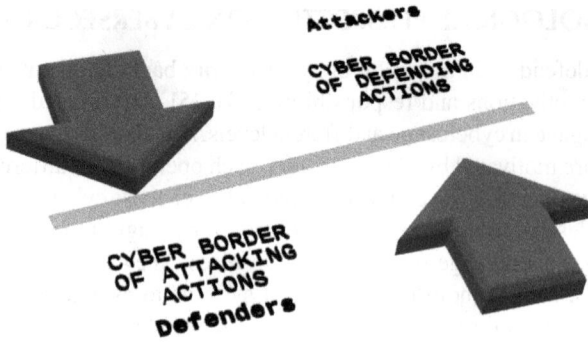

FIGURE 4.2   Cyber border of actions.

understanding the psychological aspects of cybersecurity to develop effective protection strategies in the banking sector.

In the psychological zone of cybersecurity within the banking sector, the dynamic tension between attackers and defenders can be encapsulated by the concept of a "Cyber Border of Attacking Actions" and a "Cyber Border of Defending Actions" (Figure 4.2). Attackers, driven by a variety of psychological motivations, range from individuals seeking financial gain, recognition, or revenge, to state actors engaged in espionage or economic disruption, and regional cybercrime syndicates targeting local financial institutions. These attackers operate with mindsets shaped by personal desires, political objectives, or regional ambitions, which significantly influence their methods and strategies. On the other side of this border, defenders—comprising banks, financial systems, and regular users—are guided by their psychological commitment to protect customer data, uphold trust, and ensure compliance with regulatory standards. The mental framework of these defenders involves the constant pressure to stay ahead of evolving threats, manage cybersecurity budgets effectively, and cultivate a culture of vigilance and proactive defense. The "Cyber Border of Attacking Actions" reflects the psychological drive and determination of attackers to breach defenses, while the "Cyber Border of Defending Actions" represents the resilience, adaptability, and strategic thinking required of defenders to protect digital assets. This ongoing psychological conflict not only requires technical expertise but also a profound understanding of human behavior, motivations, and the ability to anticipate and counteract adversarial psychological tactics. The impact on Industry 5.0 and Society 5.0 is significant, as these frameworks stress the integration of advanced technologies with human-centric approaches to promote sustainable development and create a super-smart society. In the realm of cybersecurity banking, Industry 5.0 encourages a synergy between AI and human intelligence to enhance security measures, while Society 5.0 leverages these advancements to address societal challenges and improve quality of life. Consequently, the psychological dimensions of cybersecurity are crucial in shaping robust systems that not only defend digital assets but also support the broader objectives of sustainable and intelligent societal progress.

4.2.1 ATTACKERS

- **Motivations**:
  - **Personal Level**: Financial gain, recognition, or revenge.

    Personal level motivations in cybersecurity are profoundly influential in shaping attacker behavior. Financial gain is a primary motivator, where individuals seek to exploit cyber vulnerabilities for monetary rewards. This pursuit often drives attackers to employ sophisticated techniques to access and steal money, financial data, or valuable assets, creating a high-stakes environment where the potential for profit can justify extensive and risky measures. Recognition is another significant factor; many attackers are propelled by the desire to achieve fame and respect within the hacker community. These individuals aim to demonstrate their technical prowess by executing high-profile attacks that garner attention and admiration from peers, seeking validation and status through their success. Revenge also plays a critical role, with some attackers targeting specific people or organizations to settle personal scores or address perceived wrongs. This motivation leads them to devise attacks that inflict disruption or harm as a form of retaliation for grievances or conflicts. Each of these personal drivers—financial incentives, the quest for recognition, and the urge for revenge—shapes how attackers approach their targets and tactics, illustrating the complex psychological factors that underpin their actions in the cybersecurity zone.
  - **National Level**: Espionage, disrupting financial stability of other nations.

    National level motivations in cybersecurity involve broader, more strategic objectives that significantly impact global security dynamics. Espionage is a key driver, where state-sponsored attackers seek to infiltrate and extract sensitive information from rival nations. This information can range from economic data to classified government communications, aiming to gain strategic advantages or influence international affairs. By compromising secure systems, these attackers gather intelligence that can be used for political maneuvering or to bolster their nation's geopolitical position. Additionally, disrupting the financial stability of other nations is another critical objective. Attacks designed to undermine a country's financial systems can destabilize economies, erode public confidence, and create broader economic repercussions. Such actions are often intended to weaken a nation's economic standing, interfere with its financial operations, or cause long-term damage to its financial infrastructure. These national-level motivations drive sophisticated and highly coordinated cyber operations, reflecting the high stakes and strategic interests involved in international cybersecurity conflicts.
  - **Regional Level**: Localized cybercrime syndicates targeting regional banks and financial systems.

    Regional level motivations in cybersecurity are characterized by localized, targeted efforts from cybercrime syndicates (or groups) that focus on specific geographic areas. These syndicates aim to exploit regional banks and financial systems, utilizing their knowledge of local infrastructure, regulations, and vulnerabilities to execute their attacks. Their primary goal is to disrupt financial operations within particular regions, steal sensitive information, or extract illicit financial gains from local institutions. By concentrating on regional targets, these cybercriminal groups can operate with a degree of precision and familiarity that enhances their ability to breach security measures and evade detection. Their activities often reflect a deep understanding of the local financial landscape, allowing them to tailor their tactics to effectively exploit weaknesses. The impact of such regional-level attacks can be significant, causing financial harm to local institutions, undermining regional economic stability, and impacting the trust and confidence of customers within those areas.

4.2.2 DEFENDERS

- **Banks and Financial Systems**:
  - **Responsibilities**: Protecting customer data, maintaining trust, and ensuring compliance with regulations.
  Banks and financial systems face critical responsibilities that profoundly influence their defensive strategies. The foremost responsibility is protecting customer data, which demands a high level of vigilance and commitment. Financial institutions must implement advanced security measures to safeguard sensitive information from unauthorized access, breaches, and theft. This involves not only deploying cutting-edge technology but also fostering a culture of security awareness among employees to prevent human errors that could compromise data integrity. Maintaining trust is another pivotal responsibility, as the confidence of customers in the security of their financial transactions is paramount. Any failure to protect data effectively can lead to a significant erosion of trust, impacting customer relationships and the institution's reputation. This aspect requires constant communication and transparency with clients to reassure them of the safety of their information and the institution's proactive stance on cybersecurity. Additionally, ensuring compliance with regulations is a critical duty that involves adhering to stringent legal standards and industry best practices designed to protect customer data and financial systems. This responsibility requires continuous monitoring and adaptation to evolving regulatory requirements, ensuring that all cybersecurity practices align with legal expectations. The psychological pressure on banks and financial systems to uphold these responsibilities is substantial, as it involves balancing the need for robust security with the imperative of maintaining customer trust and meeting regulatory obligations. This multifaceted focus shapes their overall approach to cybersecurity, driving the development and implementation of comprehensive and effective defensive strategies.

- **Challenges**: Keeping up with evolving threats and managing cybersecurity budgets.

  As cybercriminals develop increasingly sophisticated methods and tactics, financial institutions must constantly update their defenses and stay ahead of emerging risks. This involves investing in the latest technologies, conducting regular threat assessments, and adapting security protocols to address new vulnerabilities. The psychological burden of staying current with these rapid changes can be overwhelming, as it demands constant vigilance, ongoing education, and a strategic foresight to anticipate potential threats. Another critical challenge is managing cybersecurity budgets. Financial institutions must allocate resources effectively to balance between investing in advanced security technologies, training staff, and maintaining operational efficiency. Budget constraints often force institutions to make difficult decisions about where to direct their resources, potentially impacting their ability to deploy comprehensive security measures. The pressure to achieve optimal security within financial limitations can create significant stress and necessitates careful prioritization and strategic planning. Both of these challenges—keeping pace with evolving threats and managing cybersecurity budgets—require a nuanced understanding of the cybersecurity landscape and a resilient approach to maintaining robust defenses while ensuring financial prudence.

- **Regular Users**:
  - **Responsibilities**: Safeguarding personal information and practicing safe online behavior.

    Regular users face essential responsibilities that critically impact their role in defending against cyber threats. Central to their responsibilities is safeguarding personal information, which requires users to be vigilant about the security of their personal data across various digital platforms. This involves adopting secure practices such as using strong, unique passwords, avoiding sharing sensitive information through unsecured channels, and being cautious about the information they disclose online. Additionally, practicing safe online behavior is crucial; users must be proactive in recognizing and avoiding phishing attempts, malware, and other forms of social engineering designed to compromise their security. The psychological pressure on regular users stems from the need to remain constantly aware of potential threats and to adopt a disciplined approach to digital security. This responsibility demands a high level of personal discipline and awareness, as lapses in security practices can lead to significant personal and financial consequences. Users must also stay informed about the latest cybersecurity threats and best practices, which can be mentally taxing given the rapid pace of change in the cyber threat landscape. Overall, the psychological burden on regular users involves maintaining a state of vigilance and adopting proactive measures to protect their personal information, balancing the need for security with the complexities of navigating an increasingly digital world.

- **Challenges**: Lack of awareness, susceptibility to social engineering attacks.
  One significant challenge is the lack of awareness; many users are insufficiently informed about cybersecurity best practices or emerging threats. This gap in knowledge can lead to inadequate security measures, such as weak passwords or poor handling of personal data, leaving users vulnerable to attacks. The complexity of modern cybersecurity threats often overwhelms users, making it difficult for them to stay updated on the latest risks and defenses. Another critical challenge is susceptibility to social engineering attacks, where attackers manipulate users into divulging sensitive information or performing actions that compromise their security. Social engineering exploits psychological factors such as trust, fear, or urgency, tricking users into responding to phishing emails, fraudulent requests, or malicious links. The psychological impact of these challenges involves navigating an environment where the threats are not always visible or obvious, and where human error can lead to significant security breaches. Overcoming these challenges requires continuous education and heightened vigilance, as well as a proactive approach to recognizing and responding to potential social engineering tactics.

4.3 TECHNICAL PERSPECTIVE ON CYBERSECURITY

Attackers utilize a range of advanced technological tools and techniques to compromise security systems [36–41]. DDoS (Distributed Denial of Service) attacks are commonly used to overwhelm systems with excessive traffic, disrupting services and causing operational downtime. Keyloggers capture keystrokes on infected devices, enabling attackers to steal sensitive information such as usernames and passwords. Spyware is deployed to covertly monitor and collect data from user devices, providing attackers with valuable information over time. Phishing tactics involve sending deceptive emails or creating fraudulent websites to trick users into revealing personal information, exploiting human psychology and trust. All these types of attack on digital systems are a serious threat [42–54].

On the defenders' side, a comprehensive suite of technologies and strategies is employed to counter these threats, integrating HF with advanced tools to create robust defenses. Antivirus software detects and removes malicious software, providing protection against a wide array of threats. Antimalware tools offer comprehensive protection against various forms of malware, ensuring system security. Antikeylogger software prevents keyloggers from capturing keystrokes, safeguarding sensitive information. Local software for encryption ensures that data is encrypted, making it inaccessible to unauthorized users. Effective password management, involving the creation and management of strong, unique passwords, is crucial for maintaining security. Digital data security encompasses implementing robust measures to protect digital assets, including regular security audits, firewalls, and intrusion detection systems.

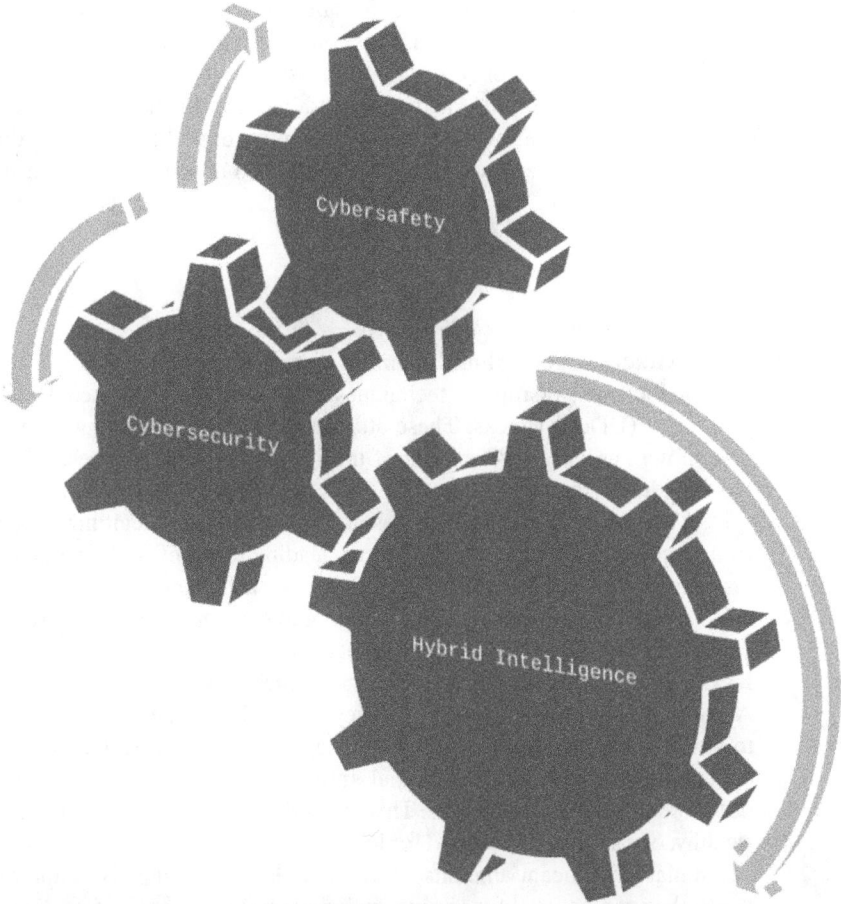

FIGURE 4.3   Hybrid intelligence covers cybersecurity and cybersafety.

Figure 4.3 illustrates how HI is essential for optimizing both cybersecurity and cybersafety, thereby ensuring the seamless operation of the entire system. HI, which merges the analytical prowess of AI with human insight and decision-making, forms the foundation of a comprehensive security framework in banking and finance operations. This integrated approach enhances the system's ability to effectively detect, analyze, and counter sophisticated cyber threats while maintaining overall system safety and integrity. In the cybersecurity, HI aids in identifying and thwarting attacks. By combining AI's rapid data processing and pattern recognition capabilities with human intuition and contextual understanding, financial institutions can respond to threats more swiftly and accurately, thereby protecting sensitive information and ensuring system stability. Cybersafety, which involves the broader protection of user data and the reliable functioning of financial systems, is also significantly bolstered by HI. Advanced technologies enhanced the synergy of AI and human oversight.

By encompassing both cybersecurity and cybersafety, HI supports the optimal performance of banking and financial systems, ensuring that all components function cohesively. This integration not only fortifies defenses against cyber threats but also maintains the safety, reliability, and trustworthiness of financial operations. Ultimately, HI is crucial for creating a resilient and adaptive digital environment that protects both the technical infrastructure and the users who rely on it, ensuring the smooth and secure operation of the entire system.

4.3.1 ATTACKERS

- **Tools and Techniques**:
  - **DDoS Attacks**: Overloading systems to disrupt services.
    One of the most disruptive techniques employed is Distributed Denial of Service (DDoS) attacks. These attacks involve overwhelming a target system with an excessive amount of traffic from multiple sources, effectively rendering the service unusable. This inundation of requests clogs the system's bandwidth, exhausts its resources, and prevents legitimate users from accessing online banking services, leading to significant operational disruptions and financial losses. The sheer scale and coordination of DDoS attacks make them particularly challenging to counter, as they can cripple critical infrastructure and erode customer trust.
  - **Keyloggers**: Capturing keystrokes to steal sensitive information.
    A particularly insidious tool is the keylogger. Keyloggers are designed to covertly capture every keystroke made on an infected device, thereby collecting a wealth of confidential information, including usernames, passwords, and financial details. This method of data theft is exceptionally stealthy, often going undetected for long periods, which allows attackers to accumulate significant amounts of sensitive data over time. The captured keystrokes can be used for various malicious purposes, such as accessing bank accounts, committing fraud, and stealing identities. The persistence and hidden nature of keyloggers make them a formidable threat, as they can bypass traditional security measures and remain active within a system indefinitely. This continuous monitoring and data collection provide attackers with a steady stream of valuable information that can be exploited for financial gain and other nefarious activities.
  - **Spyware**: Monitoring and collecting data from user devices.
    Spyware is a particularly covert tool used by cybercriminals, designed to monitor and collect data from user devices without their knowledge. This malicious software can capture a wide array of information, including browsing habits, login credentials, and personal details, all while operating silently in the background. The data gathered by spyware can be used for identity theft, financial fraud, and other malicious activities. The stealthy nature of spyware makes it especially dangerous, as it can remain undetected on a system for extended periods, continuously siphoning off valuable information. This persistent monitoring allows attackers to compile detailed profiles of their victims, which can then be used for targeted

attacks or sold on the dark web. The information captured by spyware can compromise the security of banking and financial systems, leading to significant financial losses and breaches of privacy.

- **Phishing**: Deceptive emails or websites to extract personal information.
  Phishing attacks involve sending fraudulent emails or creating counterfeit websites designed to deceive individuals into divulging sensitive personal information. These attacks often masquerade as legitimate communications from trusted entities, such as banks, financial institutions, or well-known companies, and use social engineering techniques to exploit human trust and curiosity. The fraudulent messages typically include urgent or enticing prompts, encouraging recipients to click on malicious links or provide personal data like usernames, passwords, and credit card numbers. Phishing schemes can take various forms, including email phishing, spear phishing, and pharming. Email phishing involves mass-distributing deceptive messages to a broad audience, while spear phishing targets specific individuals or organizations with tailored attacks designed to exploit particular vulnerabilities. Pharming redirects users from legitimate websites to counterfeit ones, where they are tricked into entering their credentials.

4.3.2 DEFENDERS

- **Technologies and Strategies**:
  - **Antivirus Software**: Detects and removes malicious software.
    Antivirus software is an important technology designed to detect, prevent, and remove malicious software, ensuring the protection and integrity of systems. It operates through multiple detection layers to address various threats such as viruses, worms, trojans, and ransomware. At its core, antivirus software utilizes signature-based detection to identify known malware by comparing files against a comprehensive database of threat signatures, enabling rapid pinpointing and neutralization of established threats. However, due to the rapid evolution of cyber threats, this method alone is insufficient. Modern antivirus solutions incorporate heuristic analysis to bridge this gap, evaluating file behavior and structure to identify suspicious or potentially harmful activities even if specific malware signatures are not yet known. This proactive approach allows for the detection and mitigation of emerging threats. Additionally, behavioral monitoring enables antivirus software to observe real-time program behavior, flagging deviations from normal operations and detecting malware exhibiting unusual activity patterns, which may lead to further investigation or automatic quarantine. The effectiveness of antivirus software is significantly enhanced by HI, combining AI-driven analytics with human expertise. AI technologies process and analyze large volumes of data swiftly, identifying patterns and anomalies indicative of malware, while machine learning algorithms continuously improve detection capabilities based on new threat data. Human analysts provide contextual insights, fine-tune detection algorithms, and address sophisticated threats that automated systems may not immediately

recognize, ensuring a comprehensive and adaptive defense mechanism against evolving cyber threats.

- **Antimalware**: Protects against a variety of malware threats.
  Antimalware software is designed to safeguard systems against a broad spectrum of malware threats, including worms, trojans, ransomware, adware, and spyware. It employs a multi-faceted approach to protection, starting with signature-based detection, which identifies known malware by comparing files against an extensive database of threat signatures for swift identification and removal. Modern antimalware tools also incorporate heuristic analysis to address new or unknown threats by evaluating the behavior and characteristics of files to detect potentially harmful activities that deviate from normal patterns. Behavioral monitoring further enhances security by observing real-time program behavior to flag and address suspicious activities, such as unauthorized changes to system files. Similarly to the antivirus software, the performance of antimalware software is substantially elevated through the combination of AI-powered analysis and human cybersecurity experience.
- **Antikeylogger**: Prevents keyloggers from capturing keystrokes.
  The antikeyloggers provide targeted protection to safeguard sensitive information such as passwords, financial details, and personal data. These tools utilize real-time monitoring to detect and block unauthorized processes attempting to intercept keystrokes and may employ encryption methods to obscure the data being typed. Advanced antikeyloggers also use behavioral analysis to identify and neutralize new or evolving keylogger threats that might evade traditional detection methods. The effectiveness of antikeyloggers is significantly enhanced by HI (AI & HF). AI systems analyze large datasets to detect patterns indicative of keylogger activity and adapt to new techniques, while human analysts provide contextual insights and refine detection mechanisms to address sophisticated or novel threats. This combination ensures robust protection against keyloggers, maintaining the confidentiality and security of user input.
- **Local Software for Encryption**: Encrypts data to protect against unauthorized access.
  Local software for encryption is very important for securing data by transforming it into a protected format that prevents unauthorized access. It employs complex algorithms to encrypt files, communications, and other sensitive information, making them unreadable without the correct decryption key. This local encryption ensures that even if there are physical or digital breaches, the data remains inaccessible to unauthorized individuals. The software utilizes methods such as symmetric encryption, which uses a single key for both encryption and decryption, and asymmetric encryption, which employs a pair of keys—one public and one private—to protect data. Advanced encryption tools often feature secure key management to handle encryption keys safely. The effectiveness of encryption software is significantly enhanced by HI, which combines AI & HF. AI technologies help identify anomalies and vulnerabilities in encryption practices and

adapt to emerging threats, while human analysts provide essential insights and ensure that encryption strategies are current and effective against sophisticated attacks. This synergy ensures that local encryption software provides robust protection for sensitive data, maintaining its integrity and confidentiality against unauthorized access.

- **Password Management**: Creating and managing strong, unique passwords.

  Password management is involving the creation and maintenance of strong, unique passwords. Effective tools generate complex passwords using a mix of letters, numbers, and symbols, and manage multiple passwords securely to prevent breaches. Features like automated updates and secure sharing enhance protection. HI boosts effectiveness by combining AI & HF, ensuring passwords are robust, adapting to new threats, and refining strategies to safeguard user credentials comprehensively.

- **Digital Data Security**: Implementing robust security measures to protect digital assets.

  Digital data security consists of implementing comprehensive security measures. This involves deploying a range of protective technologies and practices to secure data from unauthorized access, loss, or corruption. Robust digital data security includes encryption to protect data in transit and at rest, access controls to ensure only authorized individuals can view or modify data, and regular backups to prevent data loss in case of system failures or attacks. Additionally, security measures such as firewalls, intrusion detection systems, and secure network configurations further bolster protection against cyber threats. The effectiveness of digital data security can be enhanced by implementing experience of the HI, which merges AI-driven analytics with human oversight. AI technologies analyze vast amounts of data to detect anomalies and potential threats, while human experts provide critical context and adjust security strategies to address evolving risks.

4.4   THE ROLE OF HI IN CYBERSECURITY BANKING

The role of HI in cybersecurity is with greatest importance for shaping effective defenses and countering sophisticated cyber threats, as highlighted in Figure 4.4, which emphasizes how HI sustains cybersecurity in banking. HI merges the analytical capabilities of AI with HF, creating a powerful synergy that significantly enhances the detection, analysis, and mitigation of cyber threats. AI contributes by swiftly processing vast amounts of data, recognizing intricate patterns, and identifying anomalies that suggest potential threats, thus enabling rapid and scalable responses to evolving attack techniques. In contrast, HFs such as vigilance, decision-making, and understanding behavioral patterns are vital in complementing AI's capabilities. Human expertise ensures that technological tools are applied effectively and ethically, offering contextual insights and nuanced judgments that automated systems alone might overlook. This integration of AI-driven analytics with human oversight results in a more resilient and adaptive cybersecurity framework, improving the ability to manage the dynamic nature of modern cyber threats.

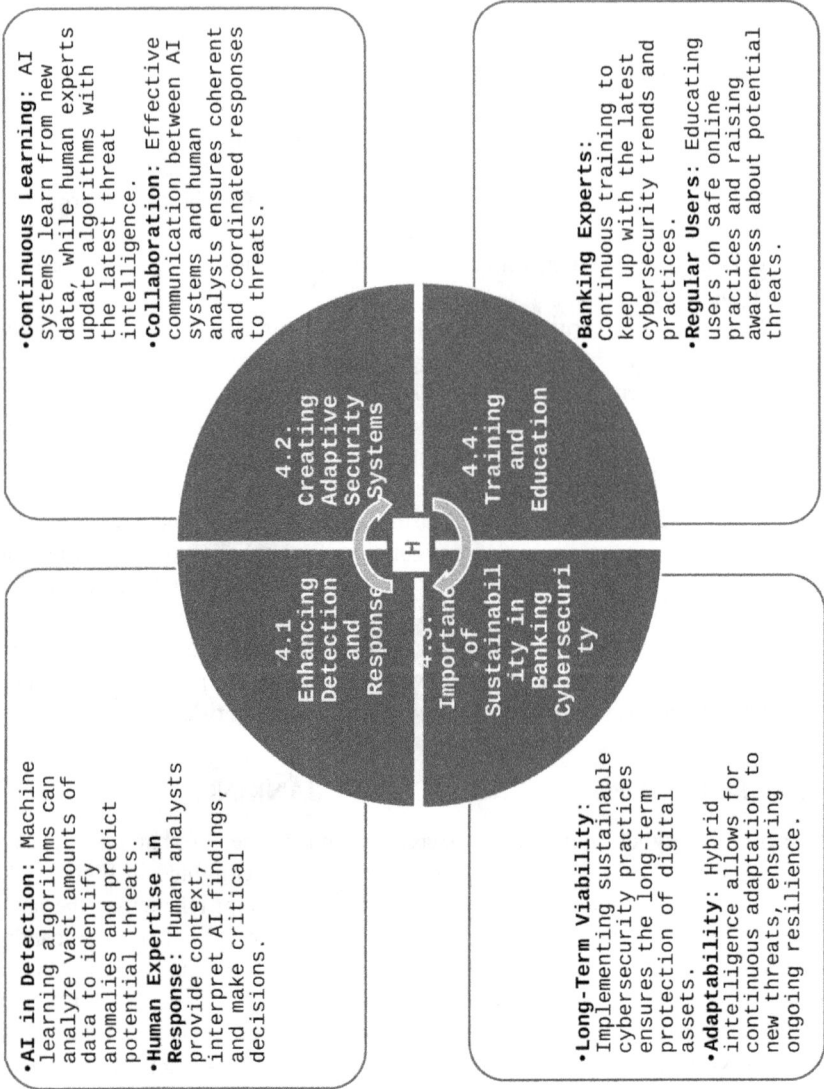

FIGURE 4.4 Hybrid intelligence (HI) sustains cybersecurity in banking.

By leveraging both advanced technology and human intelligence, HI fosters a more comprehensive and effective approach to safeguarding digital assets, supporting the principles of sustainability advocated by Society 5.0 and Industry 5.0. These frameworks emphasize creating a super-smart society and integrating human-centric approaches with technological advancements, ensuring that cybersecurity measures are not only effective but also aligned with broader goals of sustainable and human-centered development [55–67].

4.4.1 ENHANCING DETECTION AND RESPONSE

The implementation of HI in enhancing detection and response in cybersecurity is transformative, leveraging the strengths of both AI and human expertise to create a robust defense mechanism. AI significantly advances detection capabilities through machine learning algorithms that process and analyze enormous volumes of data at high speeds. These algorithms can identify subtle anomalies, detect unusual patterns, and predict potential threats with remarkable precision, thereby enabling early warning of emerging cyber risks. However, the sheer volume and complexity of data can sometimes present challenges that AI alone cannot fully address. This is where human expertise becomes indispensable. Human analysts bring critical contextual understanding and interpret the findings generated by AI systems, adding a layer of insight that is crucial for effective threat assessment and response. They are adept at making nuanced decisions based on a combination of AI outputs and their own knowledge of cybersecurity practices and potential threats. This collaborative approach ensures that AI's predictive power is complemented by human judgment, resulting in a more nuanced and effective response to cyber incidents. By integrating AI's advanced analytical capabilities with human decision-making, HI enhances both the detection of potential threats and the overall response strategy, leading to a more resilient and adaptive cybersecurity framework.

4.4.2 CREATING ADAPTIVE SECURITY SYSTEMS

Creating adaptive security systems through HI hinges on the interplay between continuous learning and effective collaboration. AI systems drive continuous learning by continuously processing new data, which allows them to recognize and adapt to emerging threats. These systems use machine learning to refine their algorithms, enhancing their ability to detect anomalies and predict potential vulnerabilities as new data flows in. Concurrently, human experts keeping these systems updated with the latest threat intelligence, ensuring that the AI remains capable of addressing the most current and sophisticated threats. The integration of human expertise ensures that algorithms are fine-tuned and aligned with real-world threat landscapes. Effective collaboration between AI and human analysts further strengthens these adaptive systems. By maintaining open and effective communication, AI-generated models are interpreted accurately, allowing for well-coordinated and timely responses to cyber threats.

4.4.3 IMPORTANCE OF SUSTAINABILITY IN BANKING CYBERSECURITY

Sustainable cybersecurity practices are essential for ensuring the enduring protection of digital assets, as they involve implementing strategies that are not only effective today but also resilient to future challenges. This approach prioritizes long-term planning and the integration of robust security measures that evolve with technological advancements and emerging threats. HI has a key role in this sustainability by enabling continuous adaptation to new threats. By combining AI's capability to analyze and respond to evolving data patterns with human expertise in threat assessment and strategic planning, HI ensures that cybersecurity systems remain resilient and capable of addressing dynamic and sophisticated attack methods. This synergy supports a proactive security posture, where systems are consistently updated and refined to mitigate risks effectively. The result is a cybersecurity framework that not only safeguards digital assets in the present but also adapts to future developments, thereby maintaining a sustainable and secure banking environment over time.

4.4.4 TRAINING AND EDUCATION

Training and education are solid components of maintaining robust cybersecurity in banking, addressing both banking experts and regular users. For banking experts, continuous training is essential to stay abreast of the latest cybersecurity trends, tools, and practices. This ongoing education helps them adapt to rapidly evolving threats, understand new security technologies, and apply best practices in their daily operations. Regular updates and training sessions ensure that professionals are well-equipped to handle emerging challenges and respond effectively to potential security incidents. Equally important is the education of regular users, who play a significant role in overall cybersecurity. Raising awareness about safe online practices and potential threats empowers users to protect their personal information and recognize phishing attempts, malware, and other common cyber threats. By promoting safe online behavior and educating users on the importance of strong passwords, secure browsing habits, and recognizing suspicious activities, banks can enhance the overall security posture. Combining the advanced training of professionals with widespread user education creates a comprehensive approach to cybersecurity, ensuring that both the technical and human elements are effectively managed to safeguard digital assets and maintain a secure banking environment.

4.5 CONCLUSION

This chapter delves into the multifaceted role of HI within cybersecurity, emphasizing its significance as both a system and a collaborative force in safeguarding digital assets. The discussion highlights how HI operates as an invisible yet crucial interaction between HFs and state-of-the-art technical tools, including advanced AI components. By integrating these elements, the chapter provides a comprehensive overview of the general concept of cybersecurity, encompassing essential terminology and exploring various methods of attack on banking and financial data. The general information underscores the importance of leveraging technological tools such as protection software, encryption, and robust password management, both by banking experts and regular users, to fortify defenses. Additionally, the chapter offers a detailed psychological perspective, examining the motivations and strategies of both attackers and defenders. This thorough examination renders the study a valuable resource for a diverse range of stakeholders, from banking professionals to everyday users. It serves as a foundational piece for future research, projects, and collaborative endeavors in the field of cybersecurity, providing insights that are crucial for advancing knowledge and practice in this ever-evolving domain.

ACKNOWLEDGMENTS

* Bulgarian Association of Ergonomics and Human Factors (BAEHF)
* The study was conducted with the support of CIII-HU-1506-01-2021 Ergonomics and Human Factors Regional Educational CEEPUS Network.

REFERENCES

[1] Chi, Der-Jang, and Zong-De Shen. 2022. "Using Hybrid Artificial Intelligence and Machine Learning Technologies for Sustainability in Going-Concern Prediction" *Sustainability* 14, no. 3: 1810. https://doi.org/10.3390/su14031810

[2] Loh, Peter K. K., Aloysius Z. Y. Lee, and Vivek Balachandran. 2024. "Towards a Hybrid Security Framework for Phishing Awareness Education and Defense" *Future Internet* 16, no. 3: 86. https://doi.org/10.3390/fi16030086

[3] Liu, Yuqi, and Zhiyong Fu. 2024. "Hybrid Intelligence: Design for Sustainable Multiverse via Integrative Cognitive Creation Model through Human–Computer Collaboration" *Applied Sciences* 14, no. 11: 4662. https://doi.org/10.3390/app14114662

[4] Mytnyk, Bohdan, Oleksandr Tkachyk, Nataliya Shakhovska, Solomiia Fedushko, and Yuriy Syerov. 2023. "Application of Artificial Intelligence for Fraudulent Banking Operations Recognition" *Big Data and Cognitive Computing* 7, no. 2: 93. https://doi.org/10.3390/bdcc7020093

[5] Pescetelli, Niccolo. 2021. "A Brief Taxonomy of Hybrid Intelligence" *Forecasting* 3, no. 3: 633–643. https://doi.org/10.3390/forecast3030039

[6] Korauš, Antonín, Eva Jančíková, Miroslav Gombár, Lucia Kurilovská, and Filip Černák. 2024. "Ensuring Financial System Sustainability: Combating Hybrid Threats through Anti-Money Laundering and Counter-Terrorist Financing Measures" *Journal of Risk and Financial Management* 17, no. 2: 55. https://doi.org/10.3390/jrfm17020055

[7] Lambrechts, Wim, Jessica S. Klaver, Lennart Koudijzer, and Janjaap Semeijn. 2021. "Human Factors Influencing the Implementation of Cobots in High Volume Distribution Centres" *Logistics* 5, no. 2: 32. https://doi.org/10.3390/logistics5020032

[8] Homayoun, Saeid, Mohammadreza Pazhohi, and Hashem Manzarzadeh Tamam. 2024. "The Effect of Innovation and Information Technology on Financial Resilience" *Sustainability* 16, no. 11: 4493. https://doi.org/10.3390/su16114493

[9] Szabó, Gyula, Balogh, Zoltán, Dovramadjiev, Tihomir, Draghici, Anca, Gajšek, Brigita, Jurčević Lulić, Tanja, Reiner, Michael, Mrugalska, Beata, Zunjic, Aleksandar. 2021. "Introducing the Ergonomics and Human Factors Regional Educational Ceepus Network" *ACTA TECHNICA NAPOCENSIS, Series: Applied Mathematics, Mechanics, and Engineering*, ISSN 1221 - 5872, pp. 201–212, Romania.

[10] Drăghici, Anca, Szabo, Gyula, Gajšek, Brigita, Mrugalska, Beata, Dovramadjiev, Tihomir and Zunjic, Aleksandar. 2022. "Ergonomics and Human Factors in the Cyber Age. The Case of Ergonomics and Human Factors Regional Educational CEEPUS Network" Timişoara: Editura Politehnica. https://doi.org/10.59168/RBWI1746

[11] Mrugalska, Beata and Dovramadjiev, Tihomir. 2022. "A Human Factors Perspective on Safety Culture" *IOS Press, Journal: Human Systems Management* 42, no. 3, pp. 299–304. https://doi.org/10.3233/HSM-220041

[12] Khuong, Nguyen Vinh, Nguyen Thi Thanh Phuong, Nguyen Thanh Liem, Cao Thi Mien Thuy, and Tran Hung Son. 2022. "Factors Affecting the Intention to Use Financial Technology among Vietnamese Youth: Research in the Time of COVID-19 and Beyond" *Economies* 10, no. 3: 57. https://doi.org/10.3390/economies10030057

[13] Mavlutova, Inese, Aivars Spilbergs, Atis Verdenhofs, Andris Natrins, Ilja Arefjevs, and Tatjana Volkova. 2023. "Digital Transformation as a Driver of the Financial Sector Sustainable Development: *An Impact on Financial Inclusion and Operational Efficiency*" *Sustainability* 15, no. 1: 207. https://doi.org/10.3390/su15010207

[14] Shiyyab, Fadi Shehab, Abdallah Bader Alzoubi, Qais Mohammad Obidat, and Hashem Alshurafat. 2023. "The Impact of Artificial Intelligence Disclosure on Financial Performance" *International Journal of Financial Studies 11*, no. 3: 115. https://doi.org/10.3390/ijfs11030115

[15] Allioui, Hanane, and Youssef Mourdi. 2023. "Exploring the Full Potentials of IoT for Better Financial Growth and Stability: A Comprehensive Survey" *Sensors* 23, no. 19: 8015. https://doi.org/10.3390/s23198015

[16] Triplett, William J. 2022. "Addressing Human Factors in Cybersecurity Leadership" *Journal of Cybersecurity and Privacy* 2, no. 3: 573–586. https://doi.org/10.3390/jcp2030029

[17] Bitkina, Olga Vl., Jaehyun Park, and Hyun K. Kim. 2022. "Measuring User-Perceived Characteristics for Banking Services: Proposing a Methodology" *International Journal of Environmental Research and Public Health* 19, no. 4: 2358. https://doi.org/10.3390/ijerph19042358

[18] Elmahgop, Faiza Omer. 2024. "Intellectual Capital and Bank Stability in Saudi Arabia: Navigating the Dynamics in a Transforming Economy" *Sustainability* 16, no. 10: 4226. https://doi.org/10.3390/su16104226

[19] D'Angelo, Chiara, Diletta Gazzaroli, Chiara Corvino, and Caterina Gozzoli. 2022. "Changes and Challenges in Human Resources Management: An Analysis of Human Resources Roles in a Bank Context (after COVID-19)" *Sustainability* 14, no. 8: 4847. https://doi.org/10.3390/su14084847

[20] Tavares, Maria C, Graça Azevedo, and Rui P. Marques. 2022. "The Challenges and Opportunities of Era 5.0 for a More Humanistic and Sustainable Society—A *Literature Review*" *Societies* 12, no. 6: 149. https://doi.org/10.3390/soc12060149

[21] Nahavandi, Saeid. 2019. "Industry 5.0—A Human-Centric Solution" *Sustainability* 11, no. 16: 4371. https://doi.org/10.3390/su11164371

[22] Mourtzis, Dimitris, John Angelopoulos, and Nikos Panopoulos. 2023. "The Future of the Human–Machine Interface (HMI) in Society 5.0" *Future Internet* 15, no. 5: 162. https://doi.org/10.3390/fi15050162

[23] Slavic, Dragana, Ugljesa Marjanovic, Nenad Medic, Nenad Simeunovic, and Slavko Rakic. 2024. "The Evaluation of Industry 5.0 Concepts: Social Network Analysis Approach" *Applied Sciences* 14, no. 3: 1291. https://doi.org/10.3390/app14031291

[24] Ziatdinov, Rushan, Madhu Sudhan Atteraya, and Rifkat Nabiyev. 2024. "The Fifth Industrial Revolution as a Transformative Step towards Society 5.0" *Societies* 14, no. 2: 19. https://doi.org/10.3390/soc14020019

[25] Alves, Joel, Tânia M. Lima, and Pedro D. Gaspar. 2023. "Is Industry 5.0 a Human-Centred Approach? A Systematic Review" *Processes* 11, no. 1: 193. https://doi.org/10.3390/pr11010193

[26] Davila-Gonzalez, Saul, and Sergio Martin. 2024. "Human Digital Twin in Industry 5.0: A Holistic Approach to Worker Safety and Well-Being through Advanced AI and Emotional Analytics" *Sensors* 24, no. 2: 655. https://doi.org/10.3390/s24020655

[27] Akundi, Aditya, Daniel Euresti, Sergio Luna, Wilma Ankobiah, Amit Lopes, and Immanuel Edinbarough. 2022. "State of Industry 5.0—Analysis and Identification of Current Research Trends" *Applied System Innovation* 5, no. 1: 27. https://doi.org/10.3390/asi5010027

[28] Hassan, Muhammad Ali, Shehnila Zardari, Muhammad Umer Farooq, Marwah M. Alansari, and Shimaa A. Nagro. 2024. "Systematic Analysis of Risks in Industry 5.0 Architecture" *Applied Sciences* 14, no. 4: 1466. https://doi.org/10.3390/app14041466

[29] Adel, Amr. 2023. "Unlocking the Future: Fostering Human–Machine Collaboration and Driving Intelligent Automation through Industry 5.0 in Smart Cities" *Smart Cities* 6, no. 5: 2742–2782. https://doi.org/10.3390/smartcities6050124

[30] Ben Salamah, Fai, Marco A. Palomino, Matthew J. Craven, Maria Papadaki, and Steven Furnell. 2023. "An Adaptive Cybersecurity Training Framework for the Education of Social Media Users at Work" *Applied Sciences* 13, no. 17: 9595. https://doi.org/10.3390/app13179595

[31] Chen, Yu-Cheng, Vincent John Mooney, III, and Santiago Grijalva. 2021. "Grid Cyber-Security Strategy in an Attacker-Defender Model" *Cryptography* 5, no. 2: 12. https://doi.org/10.3390/cryptography5020012

[32] Alnajim, Abdullah M., Shabana Habib, Muhammad Islam, Su Myat Thwin, and Faisal Alotaibi. 2023. "A Comprehensive Survey of Cybersecurity Threats, Attacks, and Effective Countermeasures in Industrial Internet of Things" *Technologies* 11, no. 6: 161. https://doi.org/10.3390/technologies11060161

[33] Almansoori, Afrah, Mostafa Al-Emran, and Khaled Shaalan. 2023. "Exploring the Frontiers of Cybersecurity Behavior: A Systematic Review of Studies and Theories" *Applied Sciences* 13, no. 9: 5700. https://doi.org/10.3390/app13095700

[34] Seid, Elias, Oliver Popov, and Fredrik Blix. 2024. "Security Attack Behavioural Pattern Analysis for Critical Service Providers" *Journal of Cybersecurity and Privacy* 4, no. 1: 55–75. https://doi.org/10.3390/jcp4010004

[35] Andrade, Roberto O., Walter Fuertes, María Cazares, Iván Ortiz-Garcés, and Gustavo Navas. 2022. "An Exploratory Study of Cognitive Sciences Applied to Cybersecurity" *Electronics* 11, no. 11: 1692. https://doi.org/10.3390/electronics11111692

[36] Ahsan, Mostofa, Kendall E. Nygard, Rahul Gomes, Md Minhaz Chowdhury, Nafiz Rifat, and Jayden F Connolly. 2022. "Cybersecurity Threats and Their Mitigation Approaches Using Machine Learning—A Review" *Journal of Cybersecurity and Privacy* 2, no. 3: 527–555. https://doi.org/10.3390/jcp2030027

[37] Saeed, Saqib, Sarah A. Suayyid, Manal S. Al-Ghamdi, Hayfa Al-Muhaisen, and Abdullah M. Almuhaideb. 2023. "A Systematic Literature Review on Cyber Threat Intelligence for Organizational Cybersecurity Resilience" *Sensors* 23, no. 16: 7273. https://doi.org/10.3390/s23167273

[38] Flor-Unda, Omar, Freddy Simbaña, Xavier Larriva-Novo, Ángel Acuña, Rolando Tipán, and Patricia Acosta-Vargas. 2023. "A Comprehensive Analysis of the Worst Cybersecurity Vulnerabilities in Latin America" *Informatics* 10, no. 3: 71. https://doi.org/10.3390/informatics10030071

[39] Grigaliūnas, Šarūnas, Rasa Brūzgienė, and Algimantas Venčkauskas. 2023. "The Method for Identifying the Scope of Cyberattack Stages in Relation to Their Impact

on Cyber-Sustainability Control over a System" *Electronics* 12, no. 3: 591. https://doi.org/10.3390/electronics1203059

[40] Kim, Kyounggon, Faisal Abdulaziz Alfouzan, and Huykang Kim. 2021. "Cyber-Attack Scoring Model Based on the Offensive Cybersecurity Framework" *Applied Sciences* 11, no. 16: 7738. https://doi.org/10.3390/app11167738

[41] Aslam, Muhammad Muzamil, Ali Tufail, Ki-Hyung Kim, Rosyzie Anna Awg Haji Mohd Apong, and Muhammad Taqi Raza. 2023. "A Comprehensive Study on Cyber Attacks in Communication Networks in Water Purification and Distribution Plants: Challenges, Vulnerabilities, and Future Prospects" *Sensors* 23, no. 18: 7999. https://doi.org/10.3390/s23187999

[42] Sadhwani, Sapna, Baranidharan Manibalan, Raja Muthalagu, and Pranav Pawar. 2023. "A Lightweight Model for DDoS Attack Detection Using Machine Learning Techniques" *Applied Sciences* 13, no. 17: 9937. https://doi.org/10.3390/app13179937

[43] Alkhudaydi, Omar Azib, Moez Krichen, and Ans D. Alghamdi. 2023. "A Deep Learning Methodology for Predicting Cybersecurity Attacks on the Internet of Things" *Information* 14, no. 10: 550. https://doi.org/10.3390/info14100550

[44] Adedeji, Kazeem B., Adnan M. Abu-Mahfouz, and Anish M. Kurien. 2023. "DDoS Attack and Detection Methods in Internet-Enabled Networks: Concept, *Research Perspectives, and Challenges" Journal of Sensor and Actuator Networks* 12, no. 4: 51. https://doi.org/10.3390/jsan12040051

[45] Lee, Changui, and Seojeong Lee. 2023. "Overcoming the DDoS Attack Vulnerability of an ISO 19847 Shipboard Data Server" *Journal of Marine Science and Engineering* 11, no. 5: 1000. https://doi.org/10.3390/jmse11051000

[46] Alahmadi, Amal A., Malak Aljabri, Fahd Alhaidari, Danyah J. Alharthi, Ghadi E. Rayani, Leena A. Marghalani, Ohoud B. Alotaibi, and Shurooq A. Bajandouh. 2023. "DDoS Attack Detection in IoT-Based Networks Using Machine Learning Models: A Survey and Research Directions" *Electronics* 12, no. 14: 3103. https://doi.org/10.3390/electronics12143103

[47] Haseeb-ur-rehman, Rana M. Abdul, Azana Hafizah Mohd Aman, Mohammad Kamrul Hasan, Khairul Akram Zainol Ariffin, Abdallah Namoun, Ali Tufail, and Ki-Hyung Kim. 2023. "High-Speed Network DDoS Attack Detection: A Survey" *Sensors* 23, no. 15: 6850. https://doi.org/10.3390/s23156850

[48] Abdullahi, Mujaheed, Yahia Baashar, Hitham Alhussian, Ayed Alwadain, Norshakirah Aziz, Luiz Fernando Capretz, and Said Jadid Abdulkadir. 2022. "Detecting Cybersecurity Attacks in Internet of Things Using Artificial Intelligence Methods: A Systematic Literature Review" *Electronics* 11, no. 2: 198. https://doi.org/10.3390/electronics11020198

[49] Mazhar, Tehseen, Hafiz Muhammad Irfan, Sunawar Khan, Inayatul Haq, Inam Ullah, Muhammad Iqbal, and Habib Hamam. 2023. "Analysis of Cyber Security Attacks and Its Solutions for the Smart grid Using Machine Learning and Blockchain Methods" *Future Internet* 15, no. 2: 83. https://doi.org/10.3390/fi15020083

[50] Aljeaid, Dania, Amal Alzhrani, Mona Alrougi, and Oroob Almalki. 2020. "Assessment of End-User Susceptibility to Cybersecurity Threats in Saudi Arabia by Simulating Phishing Attacks" *Information* 11, no. 12: 547. https://doi.org/10.3390/info11120547

[51] Fan, Zhengyang, Wanru Li, Kathryn Blackmond Laskey, and Kuo-Chu Chang. 2024. "Investigation of Phishing Susceptibility with Explainable Artificial Intelligence" *Future Internet* 16, no. 1: 31. https://doi.org/10.3390/fi16010031

[52] Eze, Chibuike Samuel, and Lior Shamir. 2024. "Analysis and Prevention of AI-Based Phishing Email Attacks" *Electronics* 13, no. 10: 1839. https://doi.org/10.3390/electronics13101839

[53] Kapan, Sibel, and Efnan Sora Gunal. 2023. "Improved Phishing Attack Detection with Machine Learning: A Comprehensive Evaluation of Classifiers and Features" *Applied Sciences* 13, no. 24: 13269. https://doi.org/10.3390/app132413269

[54] Mohamed, Gori, J. Visumathi, Miroslav Mahdal, Jose Anand, and Muniyandy Elangovan. 2022. "An Effective and Secure Mechanism for Phishing Attacks Using a Machine Learning Approach" *Processes* 10, no. 7: 1356. https://doi.org/10.3390/pr1 0071356

[55] Al-Kumaim, Nabil Hasan, and Sultan Khalifa Alshamsi. 2023. "Determinants of Cyberattack Prevention in UAE Financial Organizations: Assessing the Mediating Role of Cybersecurity Leadership" *Applied Sciences* 13, no. 10: 5839. https://doi.org/ 10.3390/app13105839

[56] Hatzivasilis, George, Sotiris Ioannidis, Michail Smyrlis, George Spanoudakis, Fulvio Frati, Ludger Goeke, Torsten Hildebrandt, George Tsakirakis, Fotis Oikonomou, George Leftheriotis, and et al. 2020. "Modern Aspects of Cyber-Security Training and Continuous Adaptation of Programmes to Trainees" *Applied Sciences* 10, no. 16: 5702. https://doi.org/10.3390/app10165702

[57] Hijji, Mohammad, and Gulzar Alam. 2022. "Cybersecurity Awareness and Training (CAT) Framework for Remote Working Employees" *Sensors* 22, no. 22: 8663. https:// doi.org/10.3390/s22228663

[58] Alrobaian, Shouq, Saif Alshahrani, and Abdulaziz Almaleh. 2023. "Cybersecurity Awareness Assessment among Trainees of the Technical and Vocational Training Corporation" *Big Data and Cognitive Computing* 7, no. 2: 73. https://doi.org/10.3390/ bdcc7020073

[59] Alterazi, Hassan A., Pravin R. Kshirsagar, Hariprasath Manoharan, Shitharth Selvarajan, Nawaf Alhebaishi, Gautam Srivastava, and Jerry Chun-Wei Lin. 2022. "Prevention of Cyber Security with the Internet of Things Using Particle Swarm Optimization" *Sensors* 22, no. 16: 6117. https://doi.org/10.3390/s22166117

[60] Villegas-Ch, William, Jaime Govea, and Iván Ortiz-Garces. 2024. "Developing a Cybersecurity Training Environment through the Integration of OpenAI and AWS" *Applied Sciences* 14, no. 2: 679. https://doi.org/10.3390/app14020679

[61] Ortiz-Ruiz, Emanuel, Juan Ramón Bermejo, Juan Antonio Sicilia, and Javier Bermejo. 2024. "Machine Learning Techniques for Cyberattack Prevention in IoT Systems: A Comparative Perspective of Cybersecurity and Cyberdefense in Colombia" *Electronics* 13, no. 5: 824. https://doi.org/10.3390/electronics13050824

[62] Alharbi, Talal, and Asifa Tassaddiq. 2021. "Assessment of Cybersecurity Awareness among Students of Majmaah University" *Big Data and Cognitive Computing* 5, no. 2: 23. https://doi.org/10.3390/bdcc5020023

[63] Aldawood, Hussain, and Geoffrey Skinner. 2019. "Reviewing Cyber Security Social Engineering Training and Awareness Programs—Pitfalls and Ongoing Issues" *Future Internet* 11, no. 3: 73. https://doi.org/10.3390/fi11030073

[64] Kuzior, Aleksandra, Paulina Brożek, Olha Kuzmenko, Hanna Yarovenko, and Tetyana Vasilyeva. 2022. "Countering Cybercrime Risks in Financial Institutions: Forecasting Information Trends" *Journal of Risk and Financial Management* 15, no. 12: 613. https://doi.org/10.3390/jrfm15120613

[65] Daah, Clement, Amna Qureshi, Irfan Awan, and Savas Konur. 2024. "Enhancing Zero Trust Models in the Financial Industry through Blockchain Integration: A Proposed Framework" *Electronics* 13, no. 5: 865. https://doi.org/10.3390/electronics13050865

[66] Rawindaran, Nisha, Liqaa Nawaf, Suaad Alarifi, Daniyal Alghazzawi, Fiona Carroll, Iyad Katib, and Chaminda Hewage. 2023. "Enhancing Cyber Security Governance and Policy for SMEs in Industry 5.0: A Comparative Study between Saudi

Arabia and the United Kingdom" *Digital* 3, no. 3: 200–231. https://doi.org/10.3390/digital3030014

[67] Mishra, Alok, Yehia Ibrahim Alzoubi, Asif Qumer Gill, and Memoona Javeria Anwar. 2022. "Cybersecurity Enterprises Policies: A Comparative Study" *Sensors* 22, no. 2: 538. https://doi.org/10.3390/s22020538.

5 Efficient Intrusion Tolerant System Based on Machine Learning and Human Behavior

Michael Ekonde Sone, Ann N. Amah, and Badrouzamani Mana

5.1 INTRODUCTION

An intrusion detection system (IDS) is a security service that monitors and analyzes system events to find and provide real-time or near real-time warning of attempts to access system resources in an unauthorized manner. Even when efficient intrusion prevention and detection techniques such as signature-based and anomaly-based are in place, an attacker will eventually break through. To circumvent this drawback for existing IDS, recent research efforts have been geared toward implementing intrusion-tolerant systems. These systems are equipped with mechanisms for reacting in real-time to an ongoing intrusion, by detecting and mitigating the attacks [1]. Intrusion tolerance within the formalism of Markov Decision Processes (MDPs) has been implemented to determine model parameters such as transition probabilities and cost in each decision stage [1], [2], [3]. To enhance the performance of intrusion-tolerant systems, the complexity of attacks due to human behavioral aspects should be considered. In Ref. [4], an efficient intrusion-tolerant system was proposed that could circumvent diverse insider threats by considering human behavioral aspects related to dispositional and situational effects. The implementation in [4] was based on the

DOI: 10.1201/9781003599142-7

MDPs which provided a mathematical framework for modeling decision-making in situations where outcomes are partly random and partly under the control of a decision-maker (human behavior). The decision-making or prediction of the outcome of each decision stage defined by the selected control was realized by associating a state transition with a probability and a cost. To enhance prediction and appropriate countermeasures, an efficient intrusion tolerant system is implemented in this chapter using a hidden Markov model (HMM). A HMM allows both observed events such as words seen in the input and hidden events such as part-of-speech tags that are taken as causal factors in the probabilistic model. Therefore, HMM in this chapter allows both observed events essentially the intrusion, and hidden events which are the human causal factors. Prediction in our efficient intrusion tolerant system will be implemented using the forward algorithm which is essentially the likelihood computation of an observation sequence given an HMM. Meanwhile, countermeasures to mitigate intrusion will be implemented using the backward algorithm which is decoding of an observation sequence and an HMM to discover the best hidden state sequence. The appropriate decoding process to be implemented will be the Viterbi decoding algorithm. The HMM will be implemented as a nonlinear two-stage convolutional code with the first stage used for the observed events and the second stage for the hidden events [5]. To improve the performance of prediction, the learning problem which is one of the three fundamental problems of an HMM will be tackled using machine learning. The transition probabilities for the observed events and the emission probabilities for the hidden events due to human behavior are trained using machine learning. Two concatenated databases are used for the training to reveal the observed and hidden states.

The complete outline of the chapter is as follows. In the next section, the review of HMM and how it could be implemented using the nonlinear convolutional code in this research will be presented. In Section 5.3, the basics of machine language and its application to behavioral cybersecurity analysis is presented. The implementation of the forward and Viterbi algorithms using the nonlinear two-stage convolutional code is presented in Section 5.4. The methodology for the training of the transition probabilities is presented in Section 5.5. Section 5.6 presents the results of the performance of the proposed model. Finally, the conclusion and future work are presented in Section 5.7.

5.2   REVIEW OF THE HIDDEN MARKOV MODEL (HMM) AND NONLINEAR CONVOLUTIONAL CODE

In this section, the review of HMM will be presented, and it will be shown how HMM could be implemented using multistage nonlinear convolutional code. In this research, a first-order HMM will be implemented as a two-stage nonlinear convolutional code.

5.2.1   REVIEW OF HMM

A HMM allows us to talk about both observed events and hidden events that are assumed to be causal factors in the probabilistic model. An HMM is specified by the following components [6]:

$Q = q_1 q_2 \dots q_N$	a set of N states
$A = a_{11} \dots a_{ij} \dots a_{NN}$	a transition probability matrix A, each ai j representing the probability of moving from the state i to state j, such that $\sum_{j=1}^{N} a_{ij} = 1 \; \forall i$
$B = b_i(O_t)$	a sequence of observation likelihoods, also called emission probabilities, each expressing the probability of an observation O_t (drawn from a vocabulary $V = v_1, v_2, \dots, v_V$) being generated from a state q_i
$\pi = \pi_1, \pi_2, \dots, \pi_N$	an initial probability distribution over states. π_i is the probability that the Markov chain will start in state i. Some states j may have $\pi_j = 0$, meaning that they cannot be initial states. Also, $\sum_{i=1}^{N} \pi_i = 1$

The different probabilities are shown graphically in Figure 5.1 [6].

From Figure 5.1, the HMM is given as input $O = O_1 O_2 \dots O_T$: a sequence of T observations, each one drawn from the vocabulary V. In this research, the vocabulary V is binary 0 and 1.

The probabilities in Figure 5.1 are defined as follows:

- The initial probabilities, $p(S_1 = i) = \pi_i$
- The transition probabilities, $p(S_t = j \mid S_{t-1} = i) = p_{ij}$
- The emission probabilities, $p(O_t = y \mid S_t = i) = q_i^y$

The above probabilities could be addressed based on the three fundamental problems that characterize the HMM, namely [6]:

- Problem 1 (Likelihood): Given an HMM $\lambda = (A, B)$ and an observation sequence O, determine the likelihood P.
- Problem 2 (Decoding): Given an observation sequence O and an HMM $\lambda = (A, B)$, discover the best-hidden state sequence Q.
- Problem 3 (Learning): Given an observation sequence O and the set of states in the HMM, learn the HMM parameters A and B.

The three problems could be solved using three different algorithms as follows:

- The likelihood computation is performed using the forward algorithm. The forward algorithm uses a table to store intermediate values as it builds up the

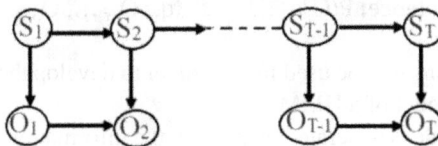

FIGURE 5.1 A first-order hidden Markov model with "hidden" variables.

probability of the observation sequence. It computes the observation probability by summing the probabilities of all possible hidden state paths that could generate the observation sequence, into a single forward trellis [6].

- The decoding problem is analyzed using the Viterbi algorithm. For any model, such as an HMM, that contains hidden variables, the task of determining which sequence of variables is the underlying source of some sequence of observations is called the decoding task. The most common decoding algorithm for HMMs is the Viterbi algorithm. Like the forward algorithm, Viterbi is a dynamic program that employs a dynamic programming trellis.

- The standard algorithm for the HMM training or the learning problem is the forward-backward algorithm or Baum- Welch algorithm [8]. Using this algorithm both the transition probabilities A and the emission probabilities B of the HMM are trained. It is an iterative algorithm, computing an initial estimate for the probabilities, then using those estimates to computing a better estimate, and so on, iteratively improving the probabilities that it learns [8].

The forward algorithm and the Viterbi will be presented in this section using a two-stage nonlinear convolutional code meanwhile the forward-backward algorithm will be presented in Section 5.3.

5.2.2 IMPLEMENTATION OF AN HMM USING A TWO-STAGE NONLINEAR CONVOLUTIONAL CODE

The forward and Viterbi algorithms of an HMM are implemented using trellis. Similarly, convolutional codes are implemented using trellis; hence, in this section, the basic concept of encoding and decoding using convolutional codes will be presented. Later, the basic convolutional code will be modified to a nonlinear convolutional code having multiple stages. In our first-order HMM using two-stage nonlinear convolutional code, the first stage will be used for observable events while the second stage will be used for hidden events.

To implement the model, the first-order HMM will instantiate two simplifying assumptions [6].

First, the probability of a particular state depends only on the previous state:

$$\textbf{Markov Assumption: } P(q_i|q_1...q_{i-1}) = P(q_i|q_{i-1}) \quad\quad (1)$$

Second, the probability of an output observation O_i depends only on the state that produced the observation q_i and not on any other states or any other observations:

$$\textbf{Output Independence: } P(O_i|q_1 \ ... \ q_i, \ ..., \ q_T, O_1, \ ..., O_i, \ ..., O_T) = P(O_i|q_i) \quad (2)$$

The two assumptions will be used in the sequel to develop the two-stage nonlinear convolutional code model of a HMM.

A convolutional code is generated by passing the information sequence to be transmitted through a linear finite-state shift register. The shift register consists of L (k-bit) stages and n linear algebraic function generators. The n linear algebraic function generators produce the n output bits for each k-bit input sequence [9], [10], [11]. The n output bits are represented by n vectors with one vector for each of the n modulo-2 adders. Hence, convolutional codes are commonly specified by three parameters; (n,k,m).

n = number of output bits
k = number of input bits
m = number of memory registers

The constraint length L which represents the number of bits in the encoder memory and affects the generation of the n output bits is defined by

$$\text{Constraint Length, } L = k\,(m-1) \tag{3}$$

Consider a (2,1,4) convolutional code as shown in Figure 5.2.

The encoding and decoding processes for the convolutional code depend on the look-up table. The look-up table consists of four items.

- Input bit
- The State of the encoder, which is one of the 8 possible states for the example (2,14) code
- The output bits. For the code (2,1,4), since 2 bits are output, the choices are 00, 01, 10, 11.
- The output state, which will be the input state for the next bit.

The look-up table for the (2,1,4) convolutional code is shown in Table 5.1.

The (2,1,4) convolutional code could be shown to implement the two assumptions of the HMM in (1) and (2) by generating the output sequence for the input sequence I = 1011. The forward trellis will involve I' = 1011000 since L zero bits are appended to the original input sequence. The first assumption will have as states $q_1 = 000_2 = 0$; $q_2 = 001_2 = 1 \ldots, q_8 = 111_2 = 7$. Meanwhile, the second assumption will have as outputs $O_1 = 00_2 = 0$; $O_2 = 01_2 = 1$; $O_3 = 10_2 = 2$; $O_4 = 11_2 = 3$.

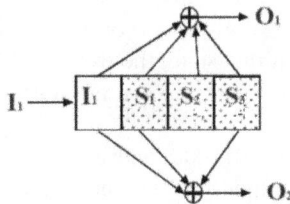

FIGURE 5.2   (2,1,4) Convolutional code.

TABLE 5.1
Look-Up Table for (2,1,4) Convolutional Code

Input Bit I_1	Input State $S_1 S_2 S_3$	Output Bits $O_1 O_2$	Output State $S_1 S_2 S_3$
0	0 0 0	0 0	0 0 0
1	0 0 0	1 1	1 0 0
0	0 0 1	1 1	0 0 0
1	0 0 1	0 0	1 0 0
0	0 1 0	1 0	0 0 1
1	0 1 0	0 1	1 0 1
0	0 1 1	0 1	0 0 1
1	0 1 1	1 0	1 0 1
0	1 0 0	1 1	0 1 0
1	1 0 0	0 0	1 1 0
0	1 0 1	0 0	0 1 0
1	1 0 1	1 1	1 1 0
0	1 1 0	0 1	0 1 1
1	1 1 0	1 0	1 1 1
0	1 1 1	1 0	0 1 1
1	1 1 1	0 1	1 1 1

5.2.2.1 Implementation of the Forward Algorithm of an HMM Using Convolutional Code

Graphically, there are three ways in which encoder operation can be represented. These are

- State diagram
- Tree diagram
- Trellis diagram

Trellis diagrams are messy but generally preferred over both the tree and the state diagrams because they represent linear time sequencing of events [9].

The Trellis diagram is drawn by lining up all the possible states (2^L) on the vertical axis. Then we connect each state to the next state by the allowable codewords for that state. There are only two choices possible at each state. These are determined by the arrival of either a 0 or a 1 bit. The diagram always begins at state 000 and starting from this state, the trellis expands and in L bits becomes fully populated such that all transitions are possible. The transitions then repeat from this point on.

Considering sequence, $I' = 1011000$, and using the look-up table in Table 5.1, the first input "1" generates output bits "11", and the current state "000" changes to

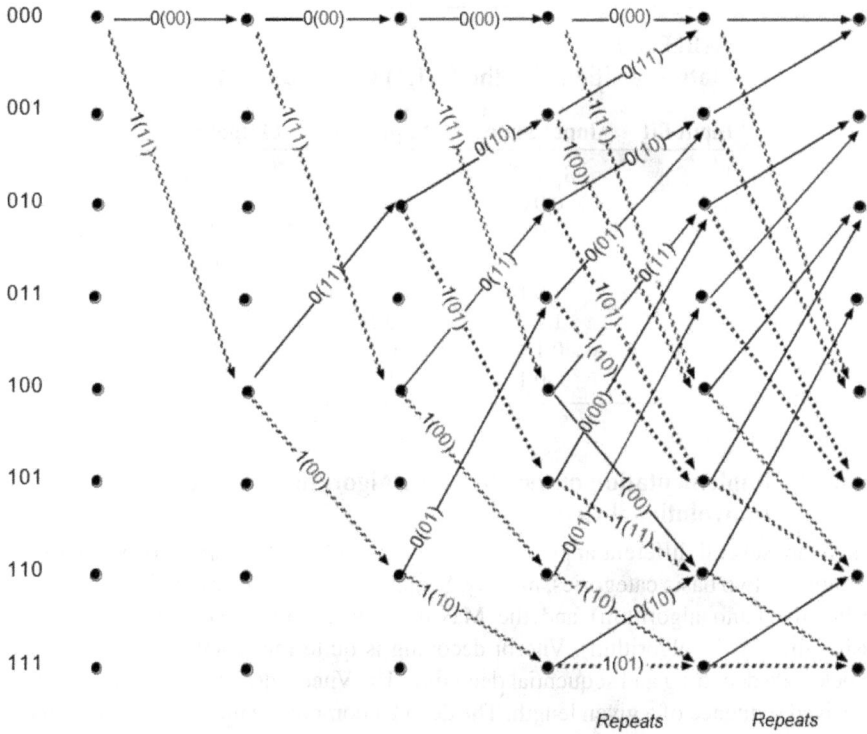

FIGURE 5.3   Trellis diagram of (2,1,4) code.

"100". The next bit "0" generates output bits "11" and the current state "100" changes to the next state "010". The entire trellis is shown in Figure 5.3 [9], [10].

The output to the complete input sequence which is observation is

$$11\ 11\ 01\ 11\ 01\ 01\ 11_2 = 3\ 3\ 1\ 3\ 1\ 1\ 3_{10}$$

In this research, instead of using the Trellis diagram as shown in Figure 5.3, a transition table derived from the states look-up table in Table 5.1 for simplicity. The transition table to obtain the observation sequence like that obtained from Figure 5.3 is shown in Table 5.2.

The final observation sequence is obtained using the output bits O_1O_2 and is given as

$$11\ 11\ 01\ 11\ 01\ 01\ 11_2 = 3\ 3\ 1\ 3\ 1\ 1\ 3_{10}$$

which is identical to the sequence obtained using the trellis in Figure 5.3.

To implement our new model of the first-order HMM, the linear (2,1,4) code will be modified into a two-stage nonlinear (2,1,4) convolutional code [5].

TABLE 5.2
State Transitions for the (2,1,4) Convolutional Code

Input Bit	Input State $S_1 S_2 S_3$	Output Bits $O_1 O_2$	Output State $S_1 S_2 S_3$
I_1			
1	0 0 0	1 1	1 0 0
0	1 0 0	1 1	0 1 0
1	0 1 0	0 1	1 0 1
1	1 0 1	1 1	1 1 0
0	1 1 0	0 1	0 1 1
0	0 1 1	0 1	0 0 1
0	0 0 1	1 1	0 0 0

5.2.2.2 Implementation of the Decoding Algorithm of an HMM Using Convolutional Code

There are several different approaches to decoding of convolutional codes. These are grouped in two basic categories, namely, Sequential Decoding (which is implemented using the Fano algorithm) and the Maximum likelihood decoding (implemented using the Viterbi algorithm). Viterbi decoding is quite important since it applies to block code decoding and sequential decoding. The Viterbi decoder examines an entire received sequence of a given length. The decoder computes a metric for each path and decides based on this metric. All paths are followed until two paths converge on one node. Then the path with the higher metric is kept and the one with the lower metric is discarded. The paths selected are called the survivors [9], [10]. This Viterbi algorithm for convolutional codes is like the Viterbi algorithm employed in the HMM which can be summarized in the following steps [10]:

- Fill out the trellis with the observation sequence from left to right.
- Compute the value of each cell $v_t(j)$ recursively by taking the most probable path that could lead to the cell. Each cell expresses the probability.

$$v_t(j) = \max_{q_1,\dots,q_{t-1}} P\left(q_1 \dots q_{t-1}, O_1, O_2 \dots O_t, q_t = j \mid \lambda\right) \qquad (4)$$

Hence, each cell of the trellis, $v_t(j)$, represents the probability that the HMM is in state j after seeing the first t observations and passing through the most probable state sequence $q_1,\dots,q_{t-1}$, given the automaton λ.

- Represent the most probable path by taking the maximum over all possible previous state sequences.

The steps above could be stated in a formal definition of the Viterbi recursion as follows [9], [10]:

- **Initialization**

$$v_1(j) = \pi_j b_j(o_1) \qquad 1 \le j \le N$$

$$b_{t_1}(j) = 0 \qquad 1 \le j \le N$$

- **Recursion**

$$v_t(j) = \max_{i=1} v_{t-1}(i) a_{ij} b_j(o_t); \quad 1 \le j \le N, \ 1 \le i \le N, \ 1 < t \le T$$

$$b_{t_1}(j) = \operatorname*{argmax}_{i=1} v_{t-1}(i) a_{ij} b_j(o_t); \quad 1 \le j \le N, \ 1 \le i \le N, \ 1 < t \le T$$

- **Termination**

The best score: $P^* = \max_{i=1} v_T(i) \qquad 1 \le i \le N$

The start of the backtrace: $q_{T^*} = \operatorname*{argmax}_{i=1} v_T(i) \qquad 1 \le i \le N$

The Viterbi algorithm will be used to find the most probable sequence of the observation sequence of the (2,1,4) convolutional code. The observation sequence which will serve as the input sequence to the decoding trellis is given as

$$11 \ 11 \ 01 \ 11 \ 01 \ 01 \ 11_2 = 3 \ 3 \ 1 \ 3 \ 1 \ 1 \ 3_{10}$$

The simplified transitions in the decoding trellis obtained using Table 5.1 are shown in Table 5.3.

TABLE 5.3
Viterbi Algorithm Design

Incoming bits	Input state	Output bits	Output state	Present metric	Cumulative metric
1 1	0 0 0	0 0	0 0 0	0	0
	0 0 0 ¹ 1 1		1 0 0	2	2
0 0	1 0 0	1 1	0 1 0	0	2
	1 0 0 ¹ 0 0		1 1 0	2	4
0 1	1 1 0	0 0 1	0 1 1	2	6
	1 1 0	1 0	1 1 1	0	4
1 0	0 1 1	0 1	0 0 1	0	6
	0 1 1 ¹ 1 0		1 0 1	2	8
0 0	1 0 1	0 0 0	0 1 0	2	10
	1 0 1	1 1	1 1 0	0	8
1 0	0 1 0	0 1 0	0 0 1	2	12
	0 1 0	0 1	1 0 1	0	10
0 0	0 0 1	1 1	0 0 0	0	12
	0 0 1 ¹ 0 0		1 0 0	2	14

The observations are referred to as the input bits in Table 5.3.

Table 5.3 shows that the best score is 13 and the backtrace starts from state transition "001" to "000".

From Table 5.3, given a sequence of observations 11 11 01 11 01 01 11_2 = 3 3 1 3 1 1 3_{10} and an HMM, the decoder fulfilled the task of finding the best-hidden sequence which is 1011_2 = 2 3_{10}.

It should be noted that the last three zeros bits are discarded since they are the constraint length, L bits.

A detailed analysis of the HMM for the two-stage nonlinear convolutional code, which considers intrusion detection and human causal factors, will be presented in Section 5.4.

5.3   BASICS OF MACHINE LEARNING AND APPLICATION TO BEHAVIORAL CYBERSECURITY

Machine learning, a subset of artificial intelligence, has gained significant traction across various disciplines due to its ability to learn from data and make predictions or decisions without being explicitly programmed. In the context of cybersecurity, machine learning plays a crucial role in enhancing threat detection and defense mechanisms by analyzing patterns and anomalies in data to identify potential security breaches. This technology allows for the development of intelligent systems that can adapt and respond to evolving cyber threats in real time.

The fundamental concepts of machine learning revolve around the idea of models and algorithms that can learn from and make predictions based on data [12]. These models are designed to optimize their performance iteratively by learning from the data they are exposed to, enabling them to improve their accuracy and efficiency over time. By utilizing machine learning techniques such as decision trees, support vector machines, neural networks, and deep learning, cybersecurity experts can create more effective models to combat complex cyber threats.

Several machine learning algorithms are categorized into three main types: supervised learning, unsupervised learning, and reinforcement learning. The choice of an algorithm depends on the nature of the data, the specific task, and the desired outcome. In the context of this research which is the implementation of an intrusion-tolerant system using machine learning and human behavior, the choice of the machine learning algorithm should be such that it can provide valuable insights into user behavior, identify patterns of malicious activity, and enhance the detection of threats both internal and external to the network [13]. The intrusion-tolerant system will be implemented as a HMM to respond to intrusions and security incidents promptly and intuitively. The model to be implemented in this research will address one of the three problems of an HMM, namely, the "Learning problem". Hence, using the K-means clustering machine learning algorithm that is widely used in cyberattack analysis in conjunction with the Forward-Backward (or the Baum-Welch algorithm) used in HMM the model will be trained to enable the analysis of vast amounts of data, identifying patterns, and enhancing threat detection and defense strategies. The model will be trained using biometric-based user authentication techniques such as facial expressions which are linked to human emotions to anticipate and understand the human behavior and traffic of malicious software [14]. This will play a critical role in preventing unauthorized access to computer systems and managing system infiltration [15]. A detailed description of the machine learning algorithm as applied to the intrusion-tolerant system will be presented in Section 5.5.

5.4 IMPLEMENTATION OF THE FORWARD AND VITERBI ALGORITHMS

In this section, the likelihood computation will be performed using the forward algorithm, and the finding of the hidden states from an observation sequence will be performed using the Viterbi algorithm for our proposed model. Our proposed model is a two-stage nonlinear (2,1,4) convolutional code which is implemented as a first-order HMM.

To implement our new model of the first-order HMM, the linear (2,1,4) code will be modified into a two-stage nonlinear convolutional code [5]. The first stage will be used for the observed events, which are generally detecting intrusion, and the second stage for the hidden events which are the human causal effects of the intrusion. The two-stage nonlinear (2,1,4) convolutional code is shown in Figure 5.4.

The transition tables of the first and second stages will be used to implement the forward and the Viterbi algorithms. The transition table for the second stage is identical to the table for the linear (2,1,4) convolutional code shown in Table 5.1 in Section 5.2. The transition table for the first stage is displayed in Table 5.4.

In the new model, the first bit is used to detect intrusion in the IDS using well-established thresholds while the next three bits are used to determine the human causal effects. In this research, we identified eight (8) different human causal effects based on different emotions such as anxiety, frustration, deception, alertness, concentration,

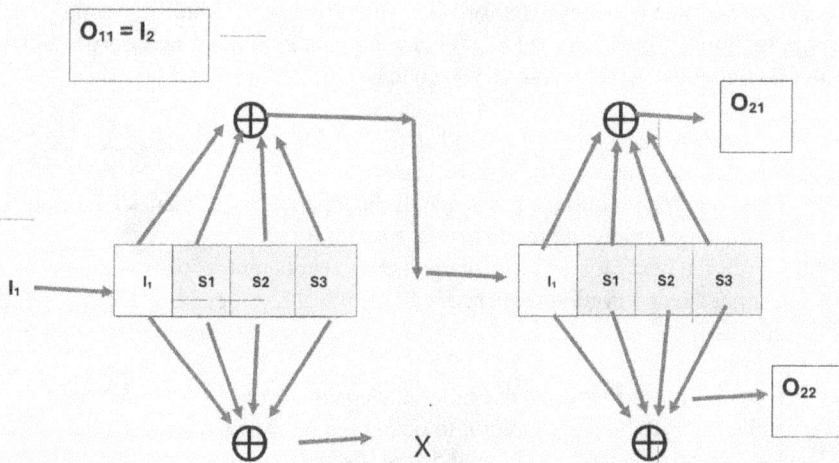

FIGURE 5.4 Two-stage nonlinear (2,1,4) convolutional code.

TABLE 5.4
Look-Up Table for First Stage of Nonlinear (2,1,4)
Convolutional Code

Incoming bits	Input state	Output bits	Output state	Present metric	Cumulative metric
1 1	0 0 0	0 0	0 0 0	0	0
	0 0 0 ¹	1 1	1 0 0	2	(2)
1 1	1 0 0	0 1 1	0 1 0	2	(4)
	1 0 0	0 0	1 1 0	0	2
0 1	0 1 0	1 0	0 0 1	0	4
	0 1 0 ¹	0 1	1 0 1	2	(6)
1 1	1 0 1	0 0	0 1 0	0	6
	1 0 1 ¹	1 1	1 1 0	2	(8)
0 1	1 1 0	0 0 1	0 1 1	2	(10)
	1 1 0	1 0	1 1 1	0	8
0 1	0 1 1	0 0 1	0 0 1	2	(12)
	0 1 1	1 0	1 0 1	0	10
1 1	0 0 1	0 1 1	0 0 0	2	(14)
	0 0 1	0 0	1 0 0	0	12

satisfaction, determination, and boredom. Once an intrusion is detected, the emotion of the intruder has to be determined to anticipate the severity level of the attack.

Hence, the following attack vectors will be used: 1000, 1001, 1010, 1011, 1100, 1101, 1110, and 1111. It should be noted that the first bit in all the attack vectors is "1" and the decision is made based on the incoming packets which are either benign or malicious. The decision for the second stage is based on another dataset which is linked to human traits.

5.4.1 IMPLEMENTATION OF THE FORWARD ALGORITHM

For each of the attack vectors, the forward algorithm will be applied, similar to the steps in section 2 for the linear (2,1,4) convolutional code. The new forward algorithm for the two-stage nonlinear (2,1,4) code has the following steps:

1. Initialize the states in stage 1 and stage 2 to "000".
2. Append the attack vector with L zeros
3. Input the first bit into the first stage. If the input bit is "1" then send the output bit of stage 1 to the input of stage 2. Else goto step 1.
4. Check the next bit and perform the likelihood computation.
5. Output the observation sequence if all the bits in the input sequence have been checked.

To illustrate the new forwarding algorithm, the attack vector, I = 1011 is used.

Append L zeros to the attack vector to obtain the input sequence, I1 = 1011000.

The look-up tables in Tables 5.1 and 5.4 are used to implement the different state transitions for the different input bits to obtain the final observation sequence. The

TABLE 5.5
State Transitions for the Two Cascaded Trellis Diagrams

	FIRST STAGE				SECOND STAGE		
Input Bit	Input State	Output Bits	Output State	Input Bit	Input State	Output Bits	Output State
I_1	$S_1 S_2 S_3$	$O_{11}=I_2$	$S_1 S_2 S_3$	I_2	$S_1 S_2 S_3$	$O_{21} O_{22}$	$S_1 S_2 S_3$
1	000	1	100	1	000	1 1	100
0	100	1	010	1	100	0 0	110
1	010	0	101	0	110	0 1	011
1	101	1	110	1	011	1 0	101
0	110	0	011	0	101	0 0	010
0	011	0	001	0	010	1 0	001
0	001	1	000	1	001	0 0	100

output sequence of the first stage is fed into the input of the second stage to generate the final observation sequence.

The simplified transitions in the trellis diagram obtained using Tables 5.1 and 5.4 are shown in Table 5.5.

The final observation sequence is obtained from the output bits in the second stage and is given as $11\ 00\ 01\ 10\ 10\ 00\ 10\ 00_2 = 3\ 0\ 1\ 2\ 2\ 0\ 2\ 0_{10}$

5.4.2 IMPLEMENTATION OF THE VITERBI ALGORITHM

In the new model, the entire process is reversed to obtain the best-hidden sequence from the observation sequence. The Viterbi algorithm is first performed for the second stage and the outputs are fed to the first stage.

The observation sequence which will serve as the input sequence to the decoding trellis of the second stage is given as

$$11\ 00\ 01\ 10\ 10\ 00\ 10\ 00_2 = 3\ 0\ 1\ 2\ 2\ 0\ 2\ 0_{10}$$

The simplified transitions in the decoding trellis obtained using Table 5.1 are shown in Table 5.6 with the observations being referred to as input bits.

The output sequence $I_2 = O_{21} = 1\ 1\ 0\ 1\ 0\ 0\ 1$ is fed into the first stage. The simplified transitions in the decoding trellis obtained using Table 5.4 are shown in Table 5.7.

From Tables 5.6 and 5.7, it is shown that given a sequence of observations $11\ 00\ 01\ 10\ 10\ 00\ 10\ 00_2 = 3\ 0\ 1\ 2\ 2\ 0\ 2\ 0_{10}$ and an HMM, the decoder fulfilled the task of finding the best-hidden sequence which is $1011_2 = 2\ 3_{10}$

It should be noted that the last three zeros bits are discarded since they are the constraint length, L bits.

TABLE 5.6
Schedule for Mapping Bits to Metric

Incoming bits	Input state	Output bits	Output state	Present metric	Cumulative metric
1 1	0 0 0	0 0	0 0 0	0	0
	0 0 0 — 1 1 1		1 0 0	2	2
0 0	1 0 0	1 1	0 10	0	2
	1 0 0 — 1 0 0		1 1 0	2	4
0 1	1 1 0 — 0 0 1		0 1 1	2	6
	1 1 0	1 0	1 1 1	0	4
1 0	0 1 1	0 1	0 0 1	0	6
	0 1 1 — 1 1 0		1 0 1	2	8
0 0	1 0 1 — 0 0 0		0 1 0	2	10
	1 0 1	1 1	1 1 0	0	8
1 0	0 1 0 — 0 1 0		0 0 1	2	12
	0 1 0	0 1	1 0 1	0	10
0 0	0 0 1	1 1	0 0 0	0	12
	0 0 1 — 1 0 0		1 0 0	2	14

TABLE 5.7
Viterbi Algorithm for the First Stage of Nonlinear (2,1,4) Code

Incoming bits	Input state	Output bits	Output state	Present metric	Cumulative metric
1	0 0 0	0	0 0 0	0	0
	0 0 0 — 1 1		1 0 0	1	1
1	1 0 0 — 0 1		0 10	1	2
	1 0 0	0	1 1 0	0	1
0	0 1 0	1	0 0 1	0	2
	0 1 0 — 1 0		1 0 1	1	3
1	1 0 1	0	0 1 0	0	3
	1 0 1 — 1 1		1 1 0	1	4
0	1 1 0 — 0 0		0 1 1	1	5
	1 1 0	1	1 1 1	0	4
0	0 1 1 — 0 0		0 0 1	1	6
	0 1 1	1	1 0 1	0	5
1	0 0 1 — 0 1		0 0 0	1	7
	0 0 1	0	1 0 0	0	6

5.5　METHODOLOGY FOR THE TRAINING OF THE MODEL

Intrusion tolerant systems (ITS) are designed to sustain operation and maintain services even during an intrusion. They achieve this by predicting the evolution of an ongoing intrusion and enabling timely defensive countermeasures. In this section the methodology used for a HMM that incorporates both machine learning techniques and human behavior analysis is discussed.

5.5.1 DATASET

The dataset used in this study is a concatenation of two well-established datasets: the Luflow dataset and the AffectNet dataset. The Luflow dataset primarily contains network traffic data classified into "malicious", "overly", and "benign" categories. The AffectNet dataset, on the other hand, provides annotated facial expressions that represent various emotional states including Anger, Contempt, Disgust, Fear, Happiness, Sadness, Surprise, and Neutral.

The decision to concatenate these datasets stems from the necessity to create a comprehensive dataset that captures both observable network traffic events and the hidden human causal factors. By combining the two datasets, the model can leverage network data for intrusion detection while using facial expression data to infer potential human factors contributing to the intrusion. This dual approach enhances the model's ability to predict and mitigate both known and unknown attacks.

5.5.2 HIDDEN MARKOV MODEL IMPLEMENTATION

The HMM is used to address the likelihood, decoding, and learning problems, fundamental to modeling intrusion tolerance systems. The model was developed and trained using the concatenated dataset, focusing on both observed and hidden events.

- Observed States: The observed data consist of two classes: "malicious" (attack) and "benign" (not attack).
- Hidden States: The hidden states are represented by the emotional states from the AffectNet dataset. These include Anger, Contempt, Disgust, Fear, Happiness, Sadness, Surprise, and Neutral.

The HMM training process involves the following steps:

1. **Data Preprocessing**: The preprocessing stage includes encoding the categorical variables from both datasets into numerical format using Label Encoding. This ensures that the data is in a suitable format for training the HMM. The concatenated dataset consisted of two categorical variables: "**label**" (representing the nature of the intrusion) and "**facial_expr**" (representing human behaviors linked to those intrusions). These variables were essential for capturing the observable and hidden elements of the probabilistic model. The dataset was loaded using the **Pandas** library, which facilitated the manipulation of data. Given that HMM requires numerical inputs, the categorical variables were encoded into numerical representations using "**LabelEncoder**" from the "**sklearn**" library.
2. **Training and Testing Split:** The concatenated dataset is split into training and testing subsets, with a typical ratio of 80:20. This split ensures that the model is trained on a substantial portion of the data while retaining a separate subset for evaluating its performance.
3. **Model Configuration and Training:** The HMM is configured with eight hidden states corresponding to the emotional states mentioned earlier.

The HMM was initialized with eight hidden states (**n_components=8**), representing different latent human behavioral patterns and their potential impact on the system's security. This configuration was selected based on the expected complexity of human factors and their influence on intrusion events.

The model is then trained using the **Baum-Welch algorithm**, which is an iterative process for estimating the parameters of the HMM that best explain the observed data. It is a standard Expectation-Maximization (EM) technique. This algorithm iteratively adjusted the HMM parameters to maximize the likelihood of the observed sequence, capturing both the direct impact of observed intrusions and the indirect effects of human factors.

The HMM's ability to infer hidden states—representing underlying human behaviors—enabled the system to make predictions about the evolution of ongoing intrusions. This predictive capability is essential for determining the most effective defensive countermeasures.

4. **Addressing HMM Fundamental Problems**
   - **Likelihood Problem**: The likelihood problem, which involves determining the probability of a sequence of observations given the model parameters, is addressed using the **Forward-Backward algorithm**. This allows for the efficient computation of the sequence probabilities.
   - **Decoding Problem:** The decoding problem focuses on determining the most likely sequence of hidden states given the observed data. The **Viterbi algorithm**, a dynamic programming approach, is employed to solve this problem, providing the optimal sequence of hidden states.
   - **Learning Problem:** The learning problem, which aims to adjust the model parameters to maximize the likelihood of the observed data, is solved using the Expectation-Maximization (EM) algorithm, specifically the **Baum-Welch** variant for HMMs. This iterative process refines the model parameters to better fit the observed data.

5. **Model Evaluation**
   The trained HMM is evaluated on the test dataset to assess its performance. The evaluation includes measuring the log-likelihood of the test data, decoding the most likely sequence of hidden states, and computing the confusion matrix and classification report. These metrics provide insights into the model's accuracy and its ability to generalize to unseen data.

   This analysis provided insights into the interaction between human factors and system vulnerabilities, helping to refine the system's predictive capabilities and enhance its resilience against both known and unknown attacks.

5.6 RESULTS AND DISCUSSION

The implementation of an intrusion-tolerant system using a HMM was evaluated based on its ability to predict intrusions and facilitate timely defensive countermeasures.

The system leveraged machine learning to address the three fundamental HMM problems: likelihood, decoding, and learning. Transition and emission matrices were derived from human behavior data to model observed and hidden events accurately.

Results

Test Log Probability and Matrices:

- **Test Log Probability**: The test log probability was calculated to be -221,516.4612985759. This value reflects the likelihood of the model generating the observed sequence of intrusions and countermeasures, indicating the model's performance on the test dataset.
- **Transition Matrix**: The learned transition matrix represents the probabilities of transitioning from one state to another in the model. The values of the matrix are as follows:

$$
\begin{bmatrix}
0.00224 & 0.00779 & 0.19725 & 0.00135 & 0.05261 & 0.00009 & 0.31052 & 0.42815 \\
0.00000 & 0.79961 & 0.00019 & 0.19978 & 0.00000 & 0.00000 & 0.00011 & 0.00030 \\
0.00911 & 0.45005 & 0.33005 & 0.01751 & 0.0000 & 0.00000 & 0.00050 & 0.19277 \\
0.00012 & 0.00000 & 0.00000 & 0.31223 & 0.00000 & 0.16583 & 0.00073 & 0.52109 \\
0.00029 & 0.00004 & 0.04398 & 0.61085 & 0.00000 & 0.19938 & 0.14067 & 0.00478 \\
0.00000 & 0.00000 & 0.00006 & 0.45525 & 0.00150 & 0.00001 & 0.00006 & 0.54313 \\
0.04170 & 0.07175 & 0.28603 & 0.30873 & 0.00023 & 0.00001 & 0.29153 & 0.00000 \\
0.00068 & 0.00000 & 0.34026 & 0.00426 & 0.00036 & 0.65197 & 0.00244 & 0.00003
\end{bmatrix}
$$

- This matrix shows the likelihood of the system transitioning from one state (column) to another (row). For instance, a high probability from state 7 to state 6 (0.65197) indicates a strong likelihood of transitioning between these states.
- **Emission Matrix**: The emission matrix outlines the probabilities of observing certain outputs from each state. The matrix is as follows:

$$
\begin{bmatrix}
0.18711 & 0.27607 & 0.10517 & 0.18515 & 0.02916 & 0.09403 & 0.000868 & 0.11464 \\
0.12459 & 0.13064 & 0.08633 & 0.12970 & 0.12970 & 0.13908 & 0.13072 & 0.12520 \\
0.15024 & 0.12479 & 0.24729 & 0.05342 & 0.14847 & 0.05014 & 0.12301 & 0.10266 \\
017548 & 0.14580 & 0.23159 & 0.10577 & 0.13021 & 0.02670 & 0.11009 & 0.07436 \\
0.00808 & 0.10408 & 0.43667 & 0.19999 & 0.05491 & 0.01286 & 0.10669 & 0.07672 \\
0.08428 & 0.11661 & 0.04601 & 0.14996 & 0.11090 & 0.18747 & 0.12623 & 0.17854 \\
0.09980 & 0.03269 & 0.18876 & 0.17446 & 0.18791 & 0.14325 & 0.10650 & 0.06663 \\
0.09780 & 0.10857 & 0.07535 & 0.14654 & 0.11510 & 0.18532 & 0.13000 & 0.14131
\end{bmatrix}
$$

This matrix indicates the likelihood of observing each of the outputs given a particular state. For example, the highest probability of observing a certain output from state 5 (0.43667) suggests a strong association between that state and the corresponding observation.

Classification Performance:

- **Confusion Matrix**: The confusion matrix revealed a significant class imbalance. It showed that all instances of the "Anger" class were misclassified as "Contempt." Specifically, out of 95,798 test instances, 47,796 belonged to the "Anger" class, and none were correctly predicted, whereas 48,002 instances were accurately classified as "Contempt."
- **Classification Report**: Reflecting the confusion matrix, the classification report indicated precision, recall, and F1-score values of zero for the "Anger" class. In contrast, the "Contempt" class had a recall of 100% and a precision of 50%. The overall accuracy of the model stood at 50%, with a macro and weighted average F1-score of 0.33.

Classification Report (Test Set)

Class	Precision	Recall	F1-Score	Support
Anger	0.00	0.00	0.00	47,796
Contempt	0.50	1,00	0.67	48,002
Class			0.50	95.798
Anger	0.25	0.50	0.33	95.798
Contempt	0.25	0.50	0.33	95.798

The dataset's distribution and the challenge of concatenating the LuFlow and AffectNet datasets significantly influence the performance of the HMM used in the intrusion-tolerant system. The balanced nature of the LuFlow dataset, with 50.19% benign and 49.81% malicious labels, provides an opportunity for the model to learn and predict both states accurately. However, the model's performance, as indicated by the confusion matrix and classification report, reveals underlying challenges that need addressing.

Concatenating the LuFlow and AffectNet datasets poses a significant challenge due to the inherent differences in their features and contexts. LuFlow focuses on network traffic with binary labels (benign or malicious), while AffectNet provides a wide range of facial expressions, each associated with distinct emotional states. This combination creates a complex dataset where the relationships between network traffic behaviors and human facial expressions are not straightforward.

The model's ability to correlate these two distinct types of data is crucial for predicting intrusions and their potential emotional motivations or impacts. However, this complexity may contribute to the misclassification observed in the confusion matrix, as the model struggles to capture the nuanced interplay between human behavior and network activity.

5.6.1 IMBALANCED LEARNING IN HMM

Despite the overall balance in the LuFlow dataset, the confusion matrix showed that all "Anger" instances were misclassified as "Contempt", indicating a discrepancy in the model's ability to handle specific facial expressions associated with different intrusion behaviors. This misclassification points to possible issues in the model's sensitivity to the nuanced differences between these expressions.

5.6.2 FACIAL EXPRESSION DISTRIBUTION

The AffectNet dataset provides in the concatenated dataset a broader spectrum of facial expressions, with "surprise" and "happy" expressions being the most prevalent, while "neutral" is the least. The proportions for "anger" (12.81%) and "contempt" (11.51%) are relatively balanced compared to other expressions, but the model's misclassification suggests that the subtle differences in expressions might not be effectively captured by the HMM.

5.6.3 IMPLICATIONS FOR INTRUSION DETECTION

The model's inability to distinguish between "anger" and "contempt" in intrusion scenarios could result in inadequate or misdirected countermeasures. Given the critical role of accurate predictions in intrusion tolerance, especially when human causal factors are involved, this limitation could have significant repercussions on the system's effectiveness.

5.6.4 RECOMMENDATIONS FOR IMPROVEMENT

To enhance the model's ability to differentiate between closely related expressions and address the challenges of concatenating diverse datasets, the following approaches could be considered:

1. **Feature Engineering**: Introducing additional features or refining the current feature set could provide the model with more discriminative power.
2. **Advanced Architectures**: Exploring more complex model architectures, such as deep learning models, may help capture the subtle differences in facial expressions and network behaviors more effectively.
3. **Data Augmentation**: Techniques to augment the existing dataset with synthetic or transformed data could help balance the representation of different facial expressions and improve the model's robustness.
4. **Class Weighting**: Implementing class weighting during model training could address the imbalanced learning and help the model focus more on the minority class.
5. **Multi-Modal Learning**: Developing a multi-modal learning framework that can effectively integrate and process the different types of data from LuFlow and AffectNet may improve the model's ability to learn the relationships between them.

5.7   CONCLUSION AND FUTURE WORK

In this chapter, a novel method to implement the HMM is presented. The method uses a two-stage nonlinear convolutional code to implement a first-order HMM. It is shown that the hidden sequence could be easily obtained from the observation sequence by using state transition tables to represent the trellis and state transition look-up tables. The novel two-stage nonlinear convolutional code is used to implement an intrusion tolerant system by detecting intrusion and the human causal effects of the detection. The intrusion is detected in the first stage of the nonlinear code while the human traits are revealed using the second stage. It was shown that the hidden sequence could be determined for any observation sequence. The three main problems in a HMM are tackled in the chapter namely, the likelihood problem, the decoding problem and the learning problem. The likelihood problem was tackled using the forward algorithm, the decoding problem was handled using the Viterbi algorithm, and the learning problem was analyzed using two cascaded datasets. While the HMM-based intrusion-tolerant system shows promise, addressing the identified challenges is crucial for enhancing its predictive capabilities and ensuring effective countermeasures for both known and unknown intrusions.

Future work will have complete training of the model to enhance prediction in the likelihood problem by determining the transition and emission probabilities. In addition, measured countermeasures will be proposed for different scenarios of intrusion based on the level of severity and the human causal effects.

REFERENCES

1. O. Patrick Kreidl, "Analysis of a Markov Decision Process Model for Intrusion Tolerance" *Dependable Systems and Networks Workshops (DSN-W), 2010 International Conference*, IEEE Xplore, August 2010.
2. B. Madan, et al. "A method for modeling and quantifying the security attributes of intrusion tolerant systems", *Performance Evaluation*, vol. 56, pp. 167–186, 2004.
3. K. Joshi, et al. "Automated recovery using bounded partially observable Markov decision processes", In *Proceedings of Dependable Systems and Networks (DSN)*, pp. 445–456, June 2006.
4. M.E. Sone, & A. Taffo, "An efficient scheme for detecting and mitigating insider threats", *New Perspectives in Behavioral Cybersecurity*, pp. 23–34, 2023, CRC Press, Boca Raton, FL. eBook.
5. M.E. Sone, "FPGA-based McEliece cryptosystem using non-linear convolutional codes", In *Proceedings of the 17th International Joint Conference on e-Business and Telecommunications*, SECRYPT 2020, ICETE (2), 2020, pp. 64–75.
6. L.R. Rabiner, "Hidden Markov model (HMM)". A tutorial on hidden Markov models and selected applications in speech recognition. *Proceedings of IEEE*, vol. 77, no. 2, pp. 257–285, February 1989.
7. H. Zen, K. Tokuda, & T. Kitamura, "Reformulating the HMM as a trajectory model by imposing explicit relationships between static and dynamic feature vector sequences". *Computer Speech and Language*, vol. 21, no. 1, pp. 153–173, 2007.
8. Z. Ghahramani, "An introduction to Hidden Markov Models and Bayesian Networks", *International Journal of Pattern Recognition and Artificial Intelligence*, vol. 15, no. 1, pp. 9–42, 2001.

9. W. Pless, *"Introduction to the Theory of Error-Correcting Codes"*, 3rd ed. New York: John Wiley & Sons, 1998.
10. C. Schlegel & L. Perez, *"Trellis Coding. Piscataway"*, NJ: IEEE Press, 1997.
11. S. B. Wicker, *"Error Control Systems for Digital Communication and Storage"*, Englewood Cliffs, NJ: Prentice Hall, 1995.
12. B. Waske, M. Fauvel, J. Benediktsson, & J. Chanussot, "Machine learning techniques in remote sensing data analysis", p. 1–24, 2009.
13. Fatai Adeshina Adelani, Enyinaya Stefano Okafor, Boma Sonimiteim Jacks, & Olakunle Abayomi Ajala, "Theoretical frameworks for the role of AI and machine learning in water cybersecurity: insights from African and U.S. applications", *Computer Science & IT Research Journal*, vol. 5, no. 3, pp. 681–692, 2024. https://doi.org/10.51594/csitrj.v5i3.928
14. D. Dasgupta, Z. Akhtar, & S. Sen, "Machine learning in cybersecurity: a comprehensive survey", *The Journal of Defense Modeling and Simulation: Applications, Methodology, Technology*, vol. 19, no. 1, pp. 57–106, 2020.
15. P. Donepudi, "Crossing point of artificial intelligence in cybersecurity", *American Journal of Trade and Policy*, vol. 2, no. 3, pp. 121–128, 2015.

Section III

Perspectives on Profiling Approaches in Cybersecurity

6 Personality Profiles of Cyber Criminals

An Integration of Existing Research

Tova Lane

6.1 INTRODUCTION

Cybercrime is a relatively young but increasingly complex area of crime, bringing continued threat to the healthy functioning of modern society (European Commission, 2020). Ranging from the ideologically motivated teenage hackers to large Cybercrime-as-a-Service (CaaS) industries, cybercriminals are as varied as the crime types themselves (Schell, 2020). An attempt to create a few singular profiles of cybercriminals would be impossible in the present cyber environment.

In an attempt to organize this relatively young field of study and threat, the European Commission, through an expert-led team, conceptualized an interdisciplinary basis for profiling of the cybercriminal (European Commission, 2022). Understanding the cybercriminal involves the integration of four different disciplines, namely, Psychology, Neuroscience, Criminology, and Cyberpsychology. A review of all of these is necessary in order to adequately profile the cybercriminal. For the purpose of this chapter, we will focus on the first of these two, namely, the psychological and neuroscientific aspects of cybercriminals.

This chapter has been written based on a non-exhaustive review of the literature on this topic. Due to the young and constantly evolving nature of this field, the reader is strongly encouraged to view all reported data with a critical and inquisitive approach. Throughout the chapter this author will repeatedly identify areas of question, and encourages the reader to do the same.

6.2 PURPOSE OF IDENTIFYING CYBER CRIMINAL PROFILES

There are many types of cybercrime and cybercriminals, each with their own hosts of motives and objectives. It is crucial to remember that due to both the complexity of the data available, as will be discussed, and dearth of data, profiles are never absolutely correct (Custers, 2021), and application of profiles for the identification of a criminal should be handled with extreme caution. Therefore, this author, as echoed by Custers (2021) strongly believes that the optimal ethical use of cybercrime profiling is to build stronger cyber intelligence for preventive measures and criminal deterrence policies.

6.3 LIMITATIONS OF THE FIELD

This brings us to the first problem in researching the field of cybercrime, the limitations of our available data. In order to understand the limitations of available research, we must begin by identifying what is needed in order to properly study cybercriminals. In order to properly research the profile of cybercriminals, we need a large quantitative dataset of cybercriminals, spread across all areas of cybercrime. We would need this data to include both criminals who have been convicted by law enforcement as well as those who have not. It would need to include criminals from all parts of the world, including parts of the world where their actions may not be illegal or legally prosecuted and may not be morally/culturally viewed as wrong. In order to properly learn about causation and development of the criminal motive and engagement over time, this data would also need to be longitudinal in nature. Additionally, due to the constantly changing nature of cybercrime, this data would need to be constantly updated to include new forms and expressions of cybercriminal behavior.

There are multiple problems in achieving such a dataset. Firstly, obtaining such data involves collaboration between numerous parties, including cybercriminals themselves who may have no incentive to aid in research targeted at understanding or preventing their activities, law enforcement agencies from whom obtaining statistically sound data can be incredibly challenging (Ugwudike, 2021), as well as victims of cybercrimes, including both individuals and corporations, who may be loath to share the data due to either lack of awareness of having been victimized, shame at having been victimized, or fear of negative implications in revealing their victimization (Kirwan 2019).

An additional dilemma when collecting data is that of taxonomy. Over the years, definitions of cybercrime and cybercriminal profiles have repeatedly changed. For example, Wall (2007) suggested three categories of cybercrime, including "computer integrity crimes", such as hacking and denial of service, "computer-assisted (or related) crimes", such as identity theft, and "computer-content crimes", such as hate crimes and child exploitation material. Another breakdown suggested by Gordon and Ford (2006) differentiated between "Type I" crimes, crimes characterized by a technological focus such as malware and fraud, as compared to "Type II" crimes, characterized as having a strong human element, such as extortion and child predation. Many additional models have been suggested as well. With this enormous variation in the taxonomy of crime, collecting uniform data becomes close to impossible.

Finally, lack of cultural considerations may render comparisons across geographic and cultural groups as statistically flawed. For example, in a small research study on Ghanan scammers (Barnor et al., 2020), the authors reviewed the spiritual meaning behind scamming. What a Western scammer might have viewed as a financially motivated crime, to the Ghanan scammers was at times a meaningful spiritual quest. Similarly, crime exists within the definitions society gives to crime (Quinney, 1970). If a corporation slips surveillance tools into their digital products, is that cybercrime? And if a hacker is acting on his or her national interests carrying out terrorist activity against another country's cyber infrastructure, is that cybercrime? Do we view that as motivated by hate? Or as motivated by a desire to achieve safety for one's family?

6.4   PSYCHOLOGY: RISK FACTORS

An early paper by Nykodym et al. (2005) described common profiles of different types of cybercriminals. In the nearly 20 years since then, the field of cybercrime has grown and developed. However, the data on risk factors remains small. When looking at risk factors, it is important to understand that different types of cybercrimes have different types of likely offenders (Schell, 2020). However, we do have some research pointing at certain observed and documented risk factors.

Aiken and colleagues published a research project in 2016 conducted on behalf of the Europol Cybercrime Centre (Aiken et al., 2016). This paper looks at interviews with expert stakeholders in the area of youth pathways into cybercrime. It was repeatedly observed that the profile with the highest risk - in their teenage years - is a male, with high IQ, high computer literacy and a strong curiosity to learn more about technology. This young man is generally somewhat socially awkward and isolated in person, but builds an online network of individuals similar to him from whom he gets social validation. This young man will generally begin with a willingness to commit small crimes, such as piracy and copyright violation, minimizing the nature of laws in the online forum. His peer group is generally one that supports this view, and as he moves from smaller to larger crimes, he receives positive reinforcement and rises through the social hierarchy of his online group. He is likely someone who derives pleasure from this increasing challenge, as well as feeding his need for affiliation and affirmation, thereby motivating him along an increasing escalation into continuously more serious crimes (Aiken et al., 2016). This prototypical young man can be found across social and economic classes.

The role of gender in understanding the cybercriminal has been studied extensively and appears to differ based on crime type, but the current research suggests an undisputed dominance of male offenders. Holt (2020) confirms this view that cybercrime, and hacking in particular, is a largely male dominated activity. Many reasons for this have been suggested, including that the anti-feminist environment and misogyny found in online forums make it less inviting for women (Barton & Laffan, 2024). Online cybercrime forums have been repeatedly identified in the literature as hotbeds of gender based sexual harassment (Barton & Laffan, 2024; Steinmetz, 2019) thus making women less likely to enter. They do not provide women with the same path of social validation as experienced by male offenders. In fact, Alfrey and Twine (2017) observe that many female members of online forums learn to actively hide their femininity in order to maintain membership in these male dominated environments. Hence, the common pathway into cybercrime, that of the teenager who finds social affiliation online, is not a pathway available to many women.

Strengthening this view, Bada et al. (2021), in reviewing a study of a large online cybercrime forum database, observed that for many male cybercriminals, their initial entry into the forum revolved around their desire to spy on their intimate partner. The nature of the discourse on these forums then remains male centric and misogynistic, at times earning the title the "manosphere" (Farrell et al., 2019).

Interestingly, the very nature of the heavily male-centric gaming world (Pan, 2023) has led to the observation of the "Queen Bee phenomenon" in which a few lead female gamers will sometimes harass other female gamers in order to defend their position in this male dominated world (Tang et al., 2020). However, this has not been extensively studied and needs clearer review.

Female cyber offenders are rarely involved in online sex crimes (Shannon, 2008; Taylor et al., 2019), but may be involved in cyberstalking or cyberbullying (Back, 2020). However, the majority of cyberbullies continue to appear to be male (Gylfason et al., 2021; van Geel et al., 2017; Wong et al., 2018), as are the majority of cyberstalkers (Dreßing et al., 2014; MacFarlane & Bocij, 2003). Additionally, sexual harassment in online video games has been documented as conducted at significantly higher rates by men than by women (Tang et al., 2020). Convicted offenders of human trafficking crimes are two thirds men and one third women, with female offenders more likely to play a role in the country of origin, frequently in Eastern Europe or Central Asia (United Nations, 2021). Crimes involving child sexual exploitation material (CSEM) appear to be predominantly engaged with by men (Quayle, 2021).

However, in a rare study looking at 98 female online sexual offenders in the US prison system, some interesting observations have been made about the production of online CSEM (Bickart et al., 2019). Bickard and colleagues found that production of online child pornography involves three parties: a motivated male, a child victim, and an individual who provides access to said child. This third party is frequently a mother providing access to her children, and playing an active role in the production of this CSEM. In further understanding the profile of this female offender, Bickart reports that they frequently have a history of significant trauma and mental health issues, and remain a criminal population needing significantly further study.

Of particular interest when trying to understand the risk factors for engaging in cybercrime, especially those related to gender, are the different predictors of cybercrime engagement for males versus females. Holt et al., (2020) reports that predictors of male cybercrime involvement are time spent with technology, while for females, peer deviance is a far stronger predictor of online criminal behavior. Additionally, while, as previously stated, men are more likely to engage in hacking (Holt et al., 2020), women are more likely to be involved in online fraud, such as using stolen credit cards to purchase goods (Lusthaus, 2018). Family support and positive relationships have also been observed to lower risk of reoffending among the female cybercriminal population in particular (Harbison, 2021), thus pointing to a different pathway and trajectory of crime for men versus women.

6.5   PSYCHOLOGY: THE DARK TETRAD

Paulhus and Williams (2002) identified three personality traits that have been termed the "modern day psychological approach to evil" (Book et al., 2016). These three traits are called the Dark Triad, and represent malicious, callous, and calculated behavior. Machiavellianism, the first of the three, is an intentional, calculated, manipulative ruthlessness toward a goal, with disregard to the harm caused to the other (Jakobwitz & Egan, 2006). Narcissism is defined as feeling superior to others and being preoccupied with a need for external admiration (Levy et al., 2007). Psychopathy is defined as impaired empathy mixed with superficial charm (Skeem et al., 2011). Later theorists (Buckels et al., 2013) suggested a Dark Tetrad constellation, theorizing that sadism, taking pleasure in inflicting pain, is a related but separate component of these antisocial personality traits.

When reviewing studies on the Dark Tetrad, it is important to distinguish between different types of cybercrimes. As previously mentioned, different crime types involve different forms of interpersonal aggression, different motivations, and hence, likely involve different offender profiles.

Beginning with cyber-dependent crimes such as hacking and ransomware, we come across a study by Siegfried-Spellar et al. (2017). Though this study, which involved self-report questionnaires administered to 235 general internet users in the US via Amazon Mechanical Turk included cyber-enabled crimes, it also looked at cyber-dependent crimes such as virus writing, website defacement, and unauthorized access (hacking). This study does not look at the entire Dark Tetrad, but narrows in on psychopathy in particular, using an assessment measure that asks questions such as "Sometimes I lie simply because I enjoy it" and "I could make a living as a con artist" (Lynam et al., 2013). Results of this study indicate that psychopathy is indeed correlated with computer crime.

Seltzer and Oelrich (2021), studying 100 German university students, found similar results. In a self-report questionnaire looking at criminal intent (such as "I have thought about hacking before" and "I have tried hacking before") and dark traits, the study authors found a positive correlation between criminal intent and machiavellianism and psychopathy. They observe that machiavellian skills such as being rational, analytic, goal oriented and strategic are all helpful traits for perpetrators of many types of cyber-dependent crime. In what will continue to be a theme in the research on dark traits, Seltzer and Oelrich (2021) did not find a correlation between narcissism and cybercrime. In an attempt to understand the reason for this, the study authors suggest that maybe cybercrime does not provide the same validation as traditional crime. This is an area for further exploration.

In continuing to review cyber-dependent crimes, Maasberg et al. (2015) proposes a model for understanding insider threat. Maasberg and colleagues propose that the Dark Triad should be an important part of understanding insider threat. Massberg suggests this based on case study reviews of well-known insider offenders identified by Band et al. (2006), such as Aldrich Ames and Robert Hanssen, and observes their reported Dark Triad traits such as chronic rule violations and lack of conscience. A weakness of this study is that of the examples mentioned, many pre-date cyber espionage and cybercrime. However, they point to features common to all insider threat attacks, thereby suggesting the importance of considering Dark Triad traits in preventing disgruntled insiders from offending.

What seems to hold up most consistently in the research on the Dark Tetrad is its correlation with cyber-enabled violence, such as trolling, cyberbullying, and cyberstalking. March (2017) studied trolling behavior on online dating apps such as Tinder and Grinder. Trolling, defined as "communication online with intention of being provocative, offensive, or menacing, in an attempt to trigger conflict and cause victims distress for the trolls own amusement" (p.140). In a study of 357 adults in Australia (71% female), self-report questionnaire results indicated that sadism and psychopathy were mostly strongly associated with cyberstalking. Interestingly, narcissism and machiavellianism were not predictive of app trolling in this study. The authors suggest that the self-absorbed nature of narcissism and the strategic manipulative style reflective of machiavellianism may simply differ from the more impulsive and disruptive nature of trolling.

In a combination of two earlier studies by Buckels et al. (2014), over 1,000 general internet users in the United States, as well as psychology students in Canada, completed voluntary self-report questionnaires. Results of these studies found sadism in particular to be positively correlated with trolling. As stated by the study authors "both trolls and sadists feel sadistic glee at the distress of others. Sadists just want to have fun... and the internet is their playground" (p. 101). Thereby pointedly describing the application of Dark Tetrad principles to the study of cyber-enabled crime.

In a self-report study of 139 adults in Iceland, trolling has been found by Gylfason et al. (2021) to be positively correlated with sadism and machiavellianism. The correlation between machiavellianism and trolling was mediated by the reported enjoyment of trolling, suggesting that "motivation to troll for Machiavellian purposes (to control and manipulate others) can therefore be understood in the context of simply taking pleasure out of the activity" (p. 8). In other words, maybe trolls who are high in machiavellian traits troll simply because they enjoy it. Of the traits studied, Gylfason found sadism to be most strongly correlated with trolling behavior, and narcissism the weakest, with a small but not statistically significant positive correlation.

An additional point that Gylfason makes regarding trolling is that trolls crave the negative social potency they get from trolling, such as enjoying making people angry or embarrassing others (Gylfason et al., 2021). Gylfason observes that this is connected with selfish and cruel behavior, reflective of the Dark Tetrad personality traits.

A study by Cracker (2016), however, takes this in a different direction. Cracker studied 396 voluntary anonymous participants on their self-reported Dark Tetrad traits and Facebook trolling behaviors. Similarly to Gylfason, Cracker observed that negative social potency is the strongest predictor of online trolling, and suggests that trolling "may be better explained by negative social reward motivation than by negative personality traits" (p. 79).

Of course, one glaring limitation when studying trolling is the potential of the trolls to troll the study questions. Self-report and self-selection in studies will always raise questions of potential limitations to accurate analysis. This may be particularly exaggerated in the case of studies of individuals prone to intentionally provide inaccurate reports of inflaming information.

Building on research looking at the connection between the Dark Tetrad and traditional (non-cyber based) bullying, studies have looked at the application of these dark traits to cyberbullying (Goodboy, 2015). In self-report of 227 college students, Goodboy (2015) found that psychopathy predicted cyberbullying. However, this study does not examine sadism. A 2017 study by van Geel and colleagues looked at 1,568 teenagers and young adults (ages 16-21 years old). This study found that sadism, but not psychopathy, is predictive of cyberbullying, and suggests that the results of Goodboy (2015) may actually reflect sadism, not psychopathy, had sadism been tested for as well (van Geel et al., 2017; Johnson et al., 2019). Van Geel concludes their analysis with the observation that it is possible that sadism, of all the Dark Tetrad traits, is the one most predictive of online antisocial behavior. They state that "online antisocial behavior then seems more driven by sadistic pleasure, than by callousness, strategic considerations, or a threatened ego" (p. 4).

Brown et al. (2019) looked further at cyberbullying and the Dark Tetrad, expanding existing research by including self-report data from 1,464 young adult participants from white, black and Asian ethnic backgrounds. Brown's study results similarly indicated a strong correlation between psychopathy and machiavellianism and cyberbullying across ethnic backgrounds. A weaker, but positive correlation between cyberbullying and both narcissism and sadism was also found.

When looking into other types of cybercrimes, the difficulty of finding primary data quickly becomes apparent. For example, there is little research on perpetrators of romance fraud and eWhoring. European Commission (2022) observes that "the callous nature of these crimes is indicative of dark personality traits; these crimes are more personal, are one-to-one attacks" (p. 83). However this author has not identified conclusive research data on the topic.

While all aforementioned studies seem to indicate at least some correlation between the so-called dark personality traits and cybercrime, it is important to observe the instances in which face value review appears to indicate situations in which this isn't the case. Understanding the exception to the rule is as important as understanding the statistically common presentation of a criminal profile.

Case study reviews point to behavior that does not align with a psychopathic view of cyber-offender traits. For example, the Ransomware-as-a-service company Darkside reportedly does not allow its affiliates to target healthcare facilities, schools, public sector, and non-profit organizations (Muncaster, 2021). Following the disruption caused by their ransomware attack on the Colonial Pipeline, they made an apology, explaining that they only meant to achieve financial gain and did not intend to cause large-scale havoc (Clark, 2021). While the application of personality traits to larger criminal organizations is not a direct one, this behavior nonetheless seems to indicate a different profile and crime motive.

In a series of qualitative interviews of romance scammers in Ghana (Barnor et al., 2020), many pointed to cultural or personal views that were not aligned with psychopathy or sadism. Some describe believing that cybercrime is less of a crime than traditional crime, indicating a lack of knowledge or awareness as to the destructive nature of the crimes involved. Others report believing that the Western victims of their scams are sufficiently wealthy that they do not lose out because of the scams. Additional research suggests that understanding neutralization, or justification of deviance, is an important aspect of understanding the cybercriminal (Brewer et al., 2020; Sykes & Matza, 1957).

When attempting to understand the contradiction in the research on psychopathy, there are several questions worth exploring. Firstly, these two aforementioned examples are case studies. More data would need to be studied in order to identify if they are simply a small exception to an otherwise accurate rule. If so, what might distinguish the exceptions from the cybercriminals matching the statistically indicated profiles? Secondly, in what ways are actions of organized crime groups, such as Darkside, significantly different from that of an individual actor? Might that alone explain the contradiction in indicated behavior versus the known research on psychopathy in cybercriminals? Furthermore, what role might culture and religious beliefs play in understanding behavior? Darkside is presumed to be a Russian based organization and forbids targeting of former Soviet states (Muncaster, 2021). Does psychopathy apply differently to actions taken against people with whom one identifies versus taken against individuals of other nationalities? Similarly, if one's culture or spiritual values encourage taking fraudulent or harmful action against another person, can that then be viewed as psychopathy? Can it still be viewed as deviant or evil? Or, as aptly stated by Goldsmith and Wall (2020) "....it must be recognized that not all motivations for transgression are indicative of deep criminal pathology or criminal career commitment" (p. 14).

When talking about terrorism, Cottee and Hayward (2011) remind us that terrorism, "for those who practice and embrace it, can be profoundly thrilling, empowering and spiritually intoxicating, and …this particular aspect of it may inform, along with other key motivations no doubt, the decision to engage in it" (p. 965). When stepping into the shoes of the terrorist, or the shoes of the romance scammer who believes his victims are not harmed, can we call that dark and evil? Is the criminal who harms another while enjoying their pain the same as the one who rationalizes or deeply believes that the victim feels no pain from his actions?

In reviewing the research on the Dark Tetrad and cyber offending, there appears to be some correlation between crimes with a human focus, such as cyberbullying and cyberstalking. The data on the correlation between these personality factors and cybercrime is less strong when considering crimes involving a technological focus and farther removed from the victim, such as hacking and malware development.

As stated at the start of this chapter, despite how young this field is, it is impossible to provide a comprehensive review of all personality factors of cybercriminals within the scope of a single chapter. Other noteworthy personality and psychological factors addressed in research include extraversion vs introversion, aggression, and tendency toward cognitive distortion. The reader is encouraged to explore these topics in the recommended readings.

6.6 NEUROSCIENCE: INTERNET ADDICTIONS AND EXCESSIVE USE

Internet addiction has generally been defined as utilization of the internet to an extent that is detrimental to the individual's daily functioning (Navarro et al., 2016). Kirwan and Power (2012) in their seminal work on the motivations of hackers, identify addiction as a possible intrinsic psychological motivator of the cybercriminal. Since then, sparse research has been done on this topic, owing partially to the lack of agreement on the construct of internet addiction (Pan et al., 2020). Aiken (2016), based on lived professional experience, states that internet addiction is an observed characteristic amongst certain profiles of cybercriminals. Some older studies, such as Nykodym et al. (2008) and Schell et al. (2020) have similarly studied this, with the most common conclusion being that it appears that heavy internet usage, but not necessarily addiction, is a predictor of cybercriminality.

In a 2015 study of 100 college students in the United States, Crimmins (2015) found that internet addiction predicted computer deviance in a majority female sample. Would these results translate into a non-college student population? Or a non-US-based population? This remains to be studied.

Of all types of cybercrime, cyberstalking has indeed been observed to be linked to internet addiction. For example, Navarro et al. (2016) surveyed over 1,600 high school students in rural North Carolina and found that the presence of an internet addiction was significantly correlated with increased rates of cyberstalking. While there appears to be some evidence suggesting a link between excessive internet use and addiction and cybercrime, the data on this is limited and inconclusive.

6.7 NEUROSCIENCE: IMPULSIVITY AND SELF-CONTROL

Researchers have suggested a link between impulsivity and self-control and cybercrime and this has been held up in some research (Palmieri et al., 2021). Additionally, childhood socialization appears to play a moderating role in this, suggesting that early childhood socialization in an environment that lacked sufficient parent guidance may be a risk factor in cyberdeviance (Schell, 2020).

It is noteworthy, however, when considering the role of impulsivity and self-control, to think about the types of crimes involved. Certain crimes, such as romance/confidence scams or complex network hacking may involve a long-term process with few short-term payoffs. These crimes therefore are less likely to involve offenders with high impulsivity. Crimes with short-term payoffs, such as trolling on dating apps, are more likely to attract individuals with high levels of impulsivity (March, 2017). However, even then, impulsivity may be moderated by other factors. For example, March (2017) found that impulsivity as a predictor of dating app trolling was moderated by medium to high levels of psychopathy. Thus, it is not the lack of self-control alone creating the inclination to commit the offense, but rather the combination of motivation co-existing with a psychopathic urge to enrage and upset other people. Hence, self-control and impulsivity should be viewed not as risk factors in and of themselves, but as a reduced deterrence in the face of strong motivation to perform the activity. The reader is recommended to read Palmieri for a further examination of the interaction between these risk factors and an introduction to criminological orientation on the topic (Palmieri et al., 2021).

6.8 DISCUSSION

Understanding human behavior involves a complex interplay of inborn temperament and biology, the socialization and environment in which one grew up, the environment in which one currently exists, and therefore the rewards and feedback one will get for their behavior within their unique environment. Criminal forums and organizations are a critical constellation of the experience of the cybercriminal and must be carefully considered in an attempt to understand the behavior and experience of the cyber offender (Holt, 2020; Lavorgna, 2020; Nurse & Bada, 2019). Cultural and ideological beliefs further shape the way in which one's behavior is internally experienced and processed. Additionally, understanding the profile of the cybercriminal involves holding awareness of a profile as a snapshot of a person who is no longer frozen in that snapshot - but an individual on a constantly moving trajectory (Horgan, 2019). Thus, motivation and presentation of the cybercriminal are constantly evolving, and doing so alongside an ever-evolving cyber and technological environment.

The reader is recommended to delve into criminological and cyberpsychology theories that provide basis and context for understanding the aforementioned psychological profile of the cybercriminal. The psychology of the individual can never be separated from the environment in which the person exists.

Finally, the reader is encouraged to consider the application of technological advances to everything that has been discussed in this chapter. As this chapter focused primarily on an individual offending actor, it has not delved into the dynamics of larger crime-as-a-service organizations and industries, and the unique characteristics that may exist amongst individuals at different levels within those organizations. Additionally, as AI continues to become an integral part of technology, it remains to be seen the ways in which this filters and shapes the motives and profiles of the individuals who engage with it for cybercriminal purposes.

RECOMMENDED FURTHER READINGS

Costa, R., Fávero, M., Moreira, D., Del Campo, A., & Sousa-Gomes, V. (2023). Dark Tetrad, acceptance of sexual violence, and sexism: A comprehensive review. *European Psychologist, 28*(1), 12–23. https://doi.org/10.1027/1016-9040/a000485

Lavorgna, A. & Holt, T. J. (2021). Researching cybercrimes: Methodologies, ethics, and critical Approaches. *Palgrave Macmillan.* https://doi.org/10.1007/978-3-030-74837-1

Martineau, M., Spiridon, E., & Aiken, M. (2023). A comprehensive framework for cyber behavioral analysis based on a systematic review of cyber profiling literature. *Forensic Science, 3,* 452–477. https://doi.org/10.3390/Forensicsci3030032

REFERENCES

Aiken, M. P. (2016). *The cyber effect*. Random House, Spiegel & Grau.

Aiken, M., Davidson, J., & Amann, P. (2016). *Youth pathways into cybercrime* (Report). Retrieved from www.europol.europa.eu/publications-documents/youth-pathways-cyb ercrime

Alfrey, L., & Twine, F. W. (2017). Gender-fluid geek girls. *Gender & Society, 31*(1), 28–50. https://doi.org/10.1177/0891243216680590

Back, S. (2019). *The cybercrime triangle*. FIU Electronic Theses and Dissertations. 4450. https://digitalcommons.fiu.edu/etd/4450

Bada, M., Chua, Y.T., Collier, B., & Pete, I. (2021). Exploring masculinities and perceptions of gender in online cybercrime subcultures. In: M. W. Kranenbarg & R. Leukfeldt, R. (Eds.), *Cybercrime in context. Crime and justice in digital society*. Springer. https://doi.org/10.1007/978-3-030-60527-8_14

Band, S. R., Cappelli, D. M., Fischer, L. F., Moore, A. P., Shaw, E. D., & Trzeciak, R. F. (2006). *Comparing insider IT sabotage and espionage: A model-based analysis*. Carnegie-Mellon University Software Engineering Institute Pittsburgh, PA, CMU/SEI-2006-TR-026.

Barnor, J. N., Boateng, R., Kolog, E. W., & Afful-Dadzie, A. (2020). Rationalizing online romance fraud: In the eyes of the offender. *AMCIS 2020 Proceedings*. 21. https://aisel.aisnet.org/amcis2020/info_security_privacy/info_security_privacy/21

Barton, H., & Laffan, D. A. (2024). The dark side of the internet. In G. Kirwan, I. Connolly, H. Barton, & M. Palmer (Eds.), *An introduction to cyberpsychology* (pp. 75–90). Routledge, Taylor & Francis Group. https://doi.org/10.4324/9781003092513

Bickart, W., Mclearen, A., Grady, M., & Stoler, K. (2019). A descriptive study of psychosocial characteristics and offense patterns in females with online child pornography offenses. *Psychiatry, Psychology and Law, 26*(2), 295–311. https://doi.org/10.1080/13218719.2018.1506714

Book, A., Visser, B. A., Blais, J., Hosker-Field, A., Methot-Jones, T., Gauthier, N. Y., Volk, A., Holden, R. R., & D'Agata, M. T. (2016). Unpacking more "evil": What is at the core of the dark tetrad? *Personality and Individual Differences, 90*, 269–272. https://doi.org/10.1016/j.paid.2015.11.009

Brewer, R., Fox, S., & Miller, C. (2020). Applying the techniques of neutralization to the study of cybercrime. In T. J. Holt & A. M. Bossler (Eds.), *The Palgrave handbook of international cybercrime and cyberdeviance* (pp. 725–742). Springer International Publishing. https://doi.org/10.1007/978-3-319-78440-3_22

Brown, W. M., Hazraty, S., & Palasinski, M. (2019). Examining the dark tetrad and its links to cyberbullying. *Cyberpsychology, Behavior and Social Networking, 22*(8), 552–557. https://doi.org/10.1089/cyber.2019.0172

Buckels, E. E., Jones, D. N., Paulhus, D. L. (2013). Behavioral confirmation of everyday sadism. *Psychological Science, 24*(11): 2201–2209. https://doi.org/10.1177/0956797613490749

Buckels, E. E., Trapnell, P. D., & Paulhus, D. L. (2014). Trolls just want to have fun. *Personality and Individual Differences, 67*, 97–102. http://dx.doi.org/10.1016/j.paid.2014.01.016.

Clark, M. (2021, May 10). *Colonial pipeline hackers apologize, promise to ransom less controversial targets in future*. The Verge. www.theverge.com/2021/5/10/22428996/colonial-pipeline-ransomware-attack-apology-investigation

Cottee, S., & Hayward, K. (2011). Terrorist (e)motives: The existential attractions of terrorism. *Studies in Conflict and Terrorism, 34*, 963–86.

Craker, N., & March, E. (2016). The dark side of Facebook®: The Dark Tetrad, negative social potency, and trolling behaviours. *Personality and Individual Differences, 102*, 79–84. https://doi .org /10 .1016 /j .paid .2016 .06 .043

Crimmins, D. M. (2015). A predictive model for self-reported computer criminal behavior among college students. *Open Access Theses.* 1210. https://docs.lib.purdue.edu/open_access_theses/1210

Custers, B. H. M. (2021). Profiling and predictions: Challenges in cybercrime research datafication. In A. Lavorgna & T. Holt & (Eds.) *Researching Cybercrimes: Methodologies, Ethics, and Critical Approaches* (pp. 63–70). Palgrave MacMillan

Dreßing, H., Bailer, J., Anders, A., Wagner, H., & Gallas, C. (2014). Cyberstalking in a large sample of social network users: Prevalence, characteristics, and impact upon victims. *Cyberpsychology, Behavior and Social Networking, 17*(2), 61–67. https://doi.org/10.1089/cyber.2012.0231

European Commission. (2020). *Communication on the EU security union strategy* (No. COM(2020) 605.

European Commission. (2022). *Drivers, trends and technology evolution in cybercrime.* https://ec.europa.eu/research/participants/documents/downloadPublic?documentIds=080166e5e93d976f&appId=PPGMS

Farrell, T., Fernandez, M., Novotny, J., & Alani, H. (2019). *Exploring misogyny across the manosphere in Reddit.* In: WebSci '19 Proceedings of the 10th ACM Conference on Web Science, pp. 87–96.

Goldsmith, A., & Wall, D. S. (2020). The seductions of cybercrime: Adolescence and the thrills of digital transgression. *European Journal of Criminology, 19*(1), 98–117. https://doi.org/10.1177/1477370819887305

Goodboy, A. K., & Martin, M. M. (2015). The personality profile of a cyberbully: Examining the Dark Triad. *Computers in Human Behavior, 49*, 1–4. https://doi.org/10.1016/j.chb.2015.02.052

Gordon, S., & Ford, R. (2006). On the definition and classification of cybercrime. *Journal in Computer Virology, 2*(1), 13–20.

Gylfason, H.F., Sveinsdottir, A.H., Vésteinsdóttir, V., & Sigurvinsdottir, R. (2021). Haters gonna hate, trolls gonna troll: The personality profile of a Facebook troll. *International Journal of Environmental Research on Public Health, 18*, 5722. https://doi.org/10.3390/Ijerph18115722

Harbison, E. (2021). Examining gender-responsive risk factors that predict recidivism for people convicted of cybercrimes. In W. M. Kranenbarg, Leukfeldt, R. (Eds.) *Cybercrime in CONTEXT. Crime and justice in digital society* (pp. 217–236). Springer. https://doi.org/10.1007/978-3-030-60527-8_14

Holt, T. J. (2020). Computer hacking and the hacker subculture. In T. J. Holt & A. M. Bossler (Eds.), *The Palgrave handbook of international cybercrime and cyberdeviance* (pp. 725–742). Springer International Publishing. https://doi.org/10.1007/978-3-319-78440-3_31 725

Holt, T. J., Navarro, J. N., & Clevenger, S. (2020). Exploring the moderating role of gender in juvenile hacking behaviors. *Crime & Delinquency, 66*(11), 1533–1555

Horgan, J. G. (2019). Psychological approaches to the study of terrorism. In E. Chenoweth, R. English, A. Gofas, & S. N. Kalyvas (Eds.), *The Oxford handbook of terrorism* (pp. 207–223). Oxford Handbooks. https://doi.org/10.1093/oxfordhb/9780198732914.013.51

Jakobwitz, S., & Egan, V. (2006). The dark triad and normal personality traits. *Personality and Individual Differences, 40*(2), 331–339. https://doi.org/10.1016/j.paid.2005.07.006

Johnson, N. F., Leahy, R., Restrepo, N. J., Velasquez, N., Zheng, M., Manrique, P., Devkota, P., & Wuchty, S. (2019). Hidden resilience and adaptive dynamics of the global online hate ecology. *Nature, 573*(7773), 261–265.

Kirwan, G., & Power, A. (2012). *The psychology of cyber crime: Concepts and principles.* Information Science Reference.

Kirwan, G. H. (2019). The rise of cybercrime. In A. Attrill-Smith, C. Fullwood, M. Keep, & D. J. Kuss (Eds.), *The Oxford handbook of cyberpsychology* (pp. 627–644). Oxford University Press.

Lavorgna, A. (2020). Organized crime and cybercrime. In T. J. Holt & A. Bossler (Eds.), *The Palgrave handbook of international cybercrime and cyberdeviance.* Palgrave Macmillan. https://doi.org/10.1007/978-3-319-78440-3_14

Lavorgna, A., & Holt, T. J. (2021). Researching cybercrimes: Methodologies, ethics, and critical approaches. *Palgrave Macmillan.* https://doi.org/10.1007/978-3-030-74837-1

Levy K.N., Reynoso J.S., Wasserman R.H., & Clarkin J.F. (2007). Narcissistic personality disorder. In W. T. O'Donohue, K. A. Fowler & S. O. Lilienfeld (Eds.), *Personality disorders: Toward the DSM-V (pp. 235).* SAGE Publications, Inc.

Lusthaus, J. (2018). *Industry of anonymity: Inside the business of cybercrime.* Harvard University Press.

Lynam, D. R., Sherman, E. D., Samuel, D., Miller, J. D., Few, L. R., & Widiger, T. A. (2013). Development of a short form of the elemental psychopathy assessment. *Assessment, 20*(6), 659–669. https://doi.org/10.1177/1073191113502072

Maasberg, M., Warren, J., & Beebe, N. L. (2015). *The dark side of the insider: Detecting the insider threat through examination of dark triad personality traits.* 48th Hawaii International Conference on System Sciences, Kauai, HI, USA, 2015, pp. 3518–3526. https://doi.org/10.1109/HICSS.2015.423.

MacFarlane, L., & Bocij, P. (2003). An exploration of predatory behaviour in cyberspace: Towards a typology of cyberstalkers. *First Monday*, 8(9).

March, E., Grieve, R., Marrington, J., & Jonason, P. K. (2017). Trolling on Tinder® (and other dating apps): Examining the role of the Dark Tetrad and impulsivity. *Personality and Individual Differences, 110*, 139–143. https://doi.org/10.1016/j.paid.2017.01.025

Muncaster, P. (2021, March 12). *Darkside 2.0 ransomware promises fastest ever encryption speeds.* Infosecurity Magazine. www.infosecurity-magazine.com/news/darkside-20-ransomware-fastest

Navarro, J. N., Marcum, C. D., Higgins, G. E., & Ricketts, M. L. (2016). Addicted to the thrill of the virtual hunt: Examining the effects of internet addiction on the cyberstalking behaviors of juveniles. *Deviant Behavior, 37*, 893–903.

Nurse, J. R. C., and Bada, M. (2019). The group element of cybercrime: Types, dynamics, and criminal operations. In A. Attrill-Smith, C. Fullwood, M. Keep, & D. J. Kuss (Eds.), *The Oxford handbook of cyberpsychology* (pp. 691–715). Oxford University Press. https://doi.org/10.1093/oxfordhb/9780198812746.013.36

Nykodym, N., Ariss, S., & Kurtz, K. (2008). Computer addiction and cyber crime. *Journal of Leadership, Accountability and Ethics. 35*, 55–59.

Palmieri, M., Shortland, N., & McGarry, P. (2021). Personality and online deviance: The role of reinforcement sensitivity theory in cybercrime. *Computers in Human Behavior, 120*, Article 106745. https://doi.org/10.1016/j.chb.2021.106745

Pan, R. (2023). Video games and gender equality: How has video gaming become a men's privilege? *Communications in Humanities Research, 6*, 37–44.

Pan, Y. C., Chiu, Y. C., & Lin, Y. H. (2020). Systematic review and meta-analysis of epidemiology of internet addiction. *Neuroscience and Biobehavioral Reviews, 118*, 612–622. https://doi.org/10.1016/j.neubiorev.2020.08.013

Paulhus, D. L., & Williams, K. M. (2002). The dark triad of personality: Narcissism, Machiavellianism, and psychopathy. *Journal of Research in Personality, 36*(6), 556–563. https://doi.org/10.1016/S0092-6566(02)00505-6

Quayle, E. (2021). Online sexual deviance and pedophilia. In L. A. Craig (Ed.), *Sexual deviance: Understanding and managing deviant sexual interests and paraphilic disorders* (pp. 222–237). Wiley.

Quinney, R. (1970). The social reality of crime. New York: Little, Brown and Company.

Schell, B. (2020). Internet addiction and cybercrime. In T. J. Holt & A. M. Bossler (Eds.), *The Palgrave handbook of international cybercrime and cyberdeviance* (pp. 679–704). Springer International Publishing. https://doi.org/10.1007/978-3-319-78440-3_26

Seigfried-Spellar, K. C., Villacís-Vukadinović, N., & Lynam, D. R. (2017). Computer criminal behavior is related to psychopathy and other antisocial behavior. *Journal of Criminal Justice, 51*, 67–73.

Selzer, N., & Oelrich, S. (2021). Saint or satan? moral development and dark triad influences on cybercriminal intent. In W. M. Kranenbarg, Leukfeldt, R. (Eds.) *Cybercrime in context. Crime and justice in digital society* (pp. 175–194). Springer. https://doi.org/10.1007/978-3-030-60527-8_14

Shannon, D. (2008). Online sexual grooming in Sweden—Online and offline sex offences against children as described in Swedish police data. *Journal of Scandinavian Studies in Criminology and Crime Prevention, 9*(2), 160–180.

Skeem, J. L., Polaschek, D. L. L., Patrick, C. J., & Lilienfeld, S. O. (2011). Psychopathic Personality. *Psychological Science in the Public Interest. 12*(3), 95–162. https://doi.org/10.1177/1529100611426706

Steinmetz, K. F., Holt, T. J., & Holt, K. M. (2019). Decoding the binary: Reconsidering the hacker subculture through a gendered lens. *Deviant Behavior, 41*(8), 936–948. https://doi.org/10.1080/01639625.2019.1596460

Sykes, G. M., & Matza, D. (1957). Techniques of neutralization: A theory of delinquency. *American Sociological Review*, 22(6), 664–670.

Tang, W. Y., Reer, F., & Quandt, T. (2020). Investigating sexual harassment in online video games: How personality and context factors are related to toxic sexual behaviors against fellow players. *Aggressive Behavior, 46*(1), 127–135. https://doi.org/10.1002/ab.21873

Taylor, R. W., Fritsch, E. J., Liederbach, J., Saylor, M. R., & Tafoya, W. L. (2019). *Cybercrime and cyber terrorism*. New York, NY: Pearson.

United Nations. (2021). *Global report on trafficking in persons*. www.unodc.org/documents/data-and-analysis/tip/2021/GLOTiP_2020_Global_overview.pdf

Van Geel, M., Goemans, A., Toprak, F., & Vedder, P. (2017). Which personality traits are related to traditional bullying and cyberbullying? A study with the Big Five, Dark Triad and sadism. *Personality and Individual Differences, 106*, 231–235.

Wall, D. S. (2007). *Cybercrime: The transformation of crime in the information age*. Cambridge: Polity Press.

Wong, R.Y., Cheung, C.M., & Xiao, B. (2018). Does gender matter in cyberbullying perpetration? An empirical investigation. *Computers in Human Behavior, 79*, 247–257. https://doi.org/10.1016/j.chb.2017.10.022

7 Cybersecurity Is Not on Maslow's Hierarchy

Implications of the Difference between Users' Physical Security and Cybersecurity Attitudes and Behaviors

Aryn Pyke

Cybersecurity is not on Maslow's Hierarchy: Implications of the difference between users' physical security and cybersecurity attitudes and behaviors

Users are not always compliant with good cybersecurity practices, and breaches are often due to human errors that cannot currently be fully mitigated by technological solutions (for an overview of human error types, see El-Bably, 2021). Some human-centric mitigations that have be suggested and attempted to improve end-user cybersecurity practices include (but are not limited to): better training on best practices, increasing awareness of consequences of bad practices; increasing ability to identify/gauge risk, and encouraging and incentivizing good cyber hygiene. However, to date one can argue that such methods have not been fully effective (e.g., Nagyfejeo & Von Solms, 2020).

The current study was motivated by the intuition that users who fail to apply good cybersecurity practices (lock computer, cover webcam, check app permissions, check website is secure), might paradoxically apply good analogous physical security

DOI: 10.1201/9781003599142-10

practices (lock doors, shut curtains, check peepholes, avoid eavesdroppers, and shred sensitive documents). If so, this raises the possibility that a more novel and effective approach to improving cybersecurity practices could involve design interventions that better harness/extend cues to activate users' existing physical security inclinations and motivations to cybersecurity.

7.1   PHYSICAL SECURITY: PRECURSOR TO CYBERSECURITY

Physical security has been important for human survival since the start of our existence, and our species has developed physically and culturally to attend to and protect ourselves and our assets from physical threats. Fischer, Halibozek, and Walters (2013, p. 3) suggested that security "implies a stable, relatively predictable environment in which an individual or group may pursue its ends without disruption or harm and without fear of disturbance or injury." In the information age, given that so many of our actions are in cyberspace, and given the prevalence of the Internet of Things, this characterization of security would seem to extend naturally to cybersecurity as well as physical security.

Furthermore, as discussed by Lee (2020), modern cybersecurity concepts and practices have often evolved from pre-existing physical security concepts and practices. In some cases, terminology for aspects of cybersecurity arise from analogies with physical security elements. For example, a firewall protecting computer networks is akin to a physical firewall in building construction, which blocks the propagation of a destructive agent, fire. Similarly, a physical door can be mechanically locked, but we can now also digitally "lock" a computer to prevent unauthorized users from getting (logging) "in".

Such parallels notwithstanding, from an evolutionary and psychological perspective, it is not clear that our instincts and motivations toward physical security transfer seamlessly to the cybersecurity domain.

7.2   SECURITY AND MASLOW'S HIERARCHY OF NEEDS

To characterize priorities in motivated behavior, Maslow proposed a 5-tier hierarchy of needs, and suggested that needs toward the base of the hierarchy would be prioritized until met, and only then would an individual be inclined/able to address higher-level needs (see Poston, 2009 for an overview). His five levels of needs in order of priority are: Physiological; Safety (i.e., security); Belongingness and Love, Esteem; and finally Self Actualization (achieving one's potential). Unsurprisingly, threads of research have explored the extent to which this hierarchy might apply to (or be harnessed to influence) behavior in cyberspace (Kellerman, 2014). For example, researchers are exploring the role of online-interactions and social media in facilitating (or not) the fulfillment of Belonginess needs (e.g., Smith, Leonis & Anandavalli, 2021).

Our current focus, however, is on security needs and behaviors. In Maslow's hierarchy, security behaviors (arising from safety needs) should be highly motivated, and second only in priority to maintaining physiological needs such as food, water, and shelter. However, pervasive non-compliance with good cybersecurity practices seems to suggest that our high motivation for security does not extend to cybersecurity. That said, people have been motivated to capitalize on cyberspace as a means to control and manage *physical security* efforts (Kellerman, 2013). For example, some harness technology and internet of thing elements like remotely accessible doorbell cameras to facilitate physical security. Pursuit of cybersecurity in its own right, however, does not seem to share the high priority/motivation that might characterize our pursuit of physical security practices.

Our degree of security motivation is undoubtedly influenced by how much we value what we are protecting. Security measures to protect our own bodies and lives (or those of our family members) would be expected to be more motivated than measures to protect digitally accessible assets (personal or banking information, intellectual property). For most of us, potential cybersecurity breaches are not mentally associated with a personal loss of life or limb. That said, in some contexts such as those involving self-driving cars and national security, such breaches can put lives at stake. Users also may also be less inclined to adopt cybersecurity practices that seem more motivated to protect their employer's assets than their own. Literally and metaphorically, users may not feel like they have much skin in the cybersecurity game. Frey (2018) suggested that ignorance and complacency in cybersecurity might largely owe to the fact that users have not experienced a major cyber incident that resulted in a prolonged loss of essential services and led to loss of life. However, from an intervention perspective, it is hoped that something short of such a drastic experience can improve risk awareness and cyber security practices.

Some physical security behaviors presumably benefit from involuntary and evolutionary processes. We may not need much/any training to eschew stagnant water, or flee, fight or freeze in the face of predators. Rather there is a readily learned or innate negative association with some cues (e.g., stagnant water, predators). Beyond needs, the key role of cues in triggering security behaviors is discussed in the next section on Situational Awareness.

7.2.1 Situational Awareness and Security Behaviors

Needs may motivate behavior, but cues in the environment also key play a role in the triggering and timing of behaviors. The key role played by environmental cues is fundamental to the concept of Situational Awareness (Endsley, 1995, 1998). Situational awareness involves perceiving cues in the environment (Observe), that signal potential threats or opportunities, comprehending their meaning, predicting what might happen next (Orient), deciding what to do (Decide), and then doing it (Act). Due to this Observe-Orient-Decide-Act iterative sequence, the situational awareness process is also known as the OODA loop, and was initially described by U.S. Air Force Colonel John Boyd in the context of fighter pilot behavior (Richards, 2020).

In everyday life, physical security threats (cues) are often proximate and concrete or at least readily envisioned (e.g., a suspicious character or car in the neighborhood, an unexpected stranger at the door, someone who seems to be starting at or following you, etc.). In contrast, it has been suggested that individuals may be less aware of cyber threats – such as a remote hypothetical hacker - because they are more abstract and out of sight, and therefore out of mind (Pyke, Feltner & Ness,

2023). Thus, despite exposing users to cautionary tales about the consequences of poor cyber hygiene, in absence of a proximate, concrete or easily envisioned threat cue, cybersecurity behaviors might not be triggered.

To raise awareness of out-of-sight cyber and electronic warfare threats in a tactical context, Pyke, Bouchelle and Uzhca (2023) exposed participants to concrete analogies for such threats via a story about a hypothetical historical mission. For example, the historical mission involved the enemy destroying physical maps (analogous to hacking or jamming a GPS) and intercepting a message transported by a carrier pigeon (analogous to intercepting wireless network or communication signals). Although the links to the corresponding modern threats were not made explicit to the participants, reading the historical mission increased their ability to identify potential cyber threats relevant to a modern mission. Thus, exposure to concrete, physical analogies presumably made the cyber threats if not literally visible, at least easier to envision and anticipate.

While encouraging, this improved awareness of cyber threats achieved via exposure to concrete physical analogies (Pyke, Bouchelle & Uzhca, 2023) was demonstrated by the participants' ability to list potential problems that might arise (Problem Anticipation Task: PAT; see also Pyke, Ness & Feltner, 2023). That task context did not require participants to be actively embedded in the situation, nor to process cues or take action to avoid or resolve issues in real-time, which would be a more ecologically valid simulation of situational awareness demands. To equip users for demands in military contexts, interfaces are being developed and refined to support decision making and timely action by presenting the user (qua military decision maker) with contextual information and real-time visible cues to represent cyber and physical threats and vulnerabilities (e.g., Llopis et al., 2018). Note that the cues presented in such common operating picture interfaces represent issues that might not otherwise be readily perceived due to such factors as their nature (e.g., electromagnetic signals), the geographic spread of the area of engagement, and the possibly distal location of the user.

For regular computer users, remote hackers/social engineers, backdoors, and computer viruses are some examples of cybersecurity threats that might not be directly perceived. Thus, it could presumably be beneficial to provide regular users with a user interface that provides salient, real-time, visible cues to represent cybersecurity threats and facilitate desired cybersecurity behaviors.

Beyond just making cybersecurity threat cues visible, an effective interface might also need to make them visceral. If, as hypothesized, users have a stronger (possibly inbuilt) inclination toward physical security than cybersecurity, then to capitalize on the affordances associated with physical security, the cyber security threat cues could resemble cues already associated with potential physical security threats (e.g., a video of someone watching you). Simulating physical threat or injury cues in a cyberspace context is not without precedent. In some first-person shooter (FPS) video games, when the security of one's avatar is compromised (e.g., takes a hit), the game interface may provide simulated visual cues associated with being physically attacked and/or injured such as displaying your (avatar's) blood spatter (e.g., Call of Duty and Battlefield), or a cracked screen effect to represent damage to your (avatar's virtual) face shield (e.g., Escape from Tarkov). Such cues might motivate actions to protect or restore the "physical" security of your avatar such as navigating away from the risky situation in virtual space or taking health power ups. In the proposed cybersecurity context, simulated physical threat cues in the user's interface could potentially be used to motivate behaviors like timely installation of security patches.

7.3   THE PRESENT RESEARCH

Poor cyber security practices might reflect a generally lax privacy and/or security mindset among young adults toward both physical and cyber security. However, it was instead hypothesized that users who fail to apply good cybersecurity practices (lock computer, cover webcam, check app permissions, check website is secure), might nonetheless apply good analogous physical security practices (lock doors, shut curtains, check peepholes, avoid eavesdroppers, and shred sensitive documents). An important implication of this hypothesis is that a novel approach to improving cybersecurity practices could involve better harnessing/extending users' extent [physical] security inclinations and motivations. To test this hypothesis, the current research investigated the relations between individuals' agreement/adherence regarding physical security versus cybersecurity practices (self-report scale: -10 = strongly disagree to 10 = strongly agree).

7.4   METHOD

7.4.1   PARTICIPANTS

College students from the United States Military Academy ($N = 130$; 33% female; mean age: 20.0 years, SD = 1.3) received course credit for participating. Note, five participants did not provide age or sex data.

7.4.2   PROCEDURE

The study was implemented on-line using Qualtrics. To get a sense of their security attitudes and actions and their expertise with and propensity to trust computers, participants completed short questionnaires, described below. Participants were also asked to provide some demographic information (e.g., sex assigned at birth, age). The presentation order of questionnaires was computer trust, security attitudes/behaviors, and then computer expertise. The whole procedure took under 30 minutes.

Physical/Cyber Security Attitudes and Behaviors. We developed these questions to gauge participants' physical security attitudes and behaviors for comparison with roughly analogous cybersecurity attitudes and behaviors. The instructions encouraged participants to be honest and reminded them that their responses would not be linked to their name. Table 7.1 summarizes the questions in the physical security and cybersecurity subscales, aligning and numbering the items to reflect roughly analogous items across the two subscales. However, this correspondence does not apply for the two rows shaded in grey. Rather, physical security item 7 (Shred sensitive information) intuitively may be related to item 6 [Avoid eavesdroppers]. The other grey row includes an extra cybersecurity question, item 8, that had a binary yes/no response about locking one's cell phone. Responses for items 1-7 were made using a slider-scale ranging from strongly disagree (-10) to strongly agree (10). The bold label in square brackets at the start of each question in Table 7.1 was not shown to participants but is used in this chapter as a convenient shorthand to later reference the items. The random presentation order of physical security items was 7, 1, 4, 5, 3, 2, 6, and random presentation order of cybersecurity items was 7, 3, 1, 6, 5, 2, 4, 8.

Propensity to Trust Computers. This 6-item scale was adapted from a Propensity to Trust Technology scale (Merrit, Heimbaugh, LaChapell, & Lee, 2013). This scale about trust in computers was included to see if there was support for the hypothesis that the lower an individual's propensity to trust computers, the higher might be his/her/their motivation to gain computer expertise and adopt strong cyber security practices. Participants responded to each item using a slider scale ranging from strongly disagree (-10) to strongly agree (10). The items are presented in Table 7.2.

Computer Expertise. For this 8-item self-report survey on a user's computer expertise, again participants responded to each item using a slider scale ranging from strongly disagree (-10) to strongly agree (10). The items are reported in Table 7.3.

7.5   RESULTS

Results are for N=130 participants. Prior to aggregating responses within a scale, where necessary, item responses were reverse coded to align with the direction of the other items (i.e., stronger agreement reflecting more stringent security practices, or greater expertise, or greater trust). In Tables 7.2, 7.3, and 7.4 items requiring reverse coding are shown in red for expository purposes, however all items were presented in black font to participants.

7.5.1   COMPUTER TRUST AND COMPUTER EXPERTISE

The main focus of the current research was on the relationship between physical and cyber security attitudes and behaviors, however, in this section we summarize the results for the other two scales: Propensity to Trust Computers (Table 7.2) and Computer Expertise (Table 7.3). In each table, the final row is the aggregation (mean) of responses within that scale (with reverse coding of items where necessary). Overall, for these scales, participants' propensity to trust computers (M=4.4), was higher than their self-reported computer expertise (M=-.01), based on responses made on a slider scale from strongly disagree = -10 to strongly agree=10. Thus, users generally tend to trust computers/programs in spite of (or perhaps out of necessity because of) their relative lack of computer expertise.

7.5.2   PHYSICAL SECURITY VERSUS CYBERSECURITY ATTITUDES AND BEHAVIORS

Table 7.4 in this section summarizes: (i) percent of participants who agreed at all (response > 0) and somewhat strongly (response>5); (ii) descriptive statistics (M, SD) for individual items in the physical and cyber security subscales; (iii) t-test

TABLE 7.1
Physical Security and Cybersecurity Items Aligned to Show the Rough Correspondence between Items in the Two Subscales

Item	Physical Security Subscale	Cybersecurity Subscale
1	**[Lock door]** If I lived in my own house or apartment [not on campus], I'd always ensure the doors/windows were locked when leaving.	**[Lock Computer]** I sign out of my account and/or remove my CAC card [ID card providing access to certain sites/services] whenever I leave my computer unattended.
	[Spare key] For my own house/ apartment [off campus], I'd have a key hidden near the entrance in case I lost/forgot my keys.	**[Re-use passwords]** I use the same password for multiple apps/ websites.
2	**[Close curtains]** In my own house, I'd usually keep my bedroom and/or other curtains facing the road or facing neighbor's windows closed for privacy.	**[Cover webcam]** I usually have a sticker or something covering my webcam/laptop camera unless I am using it.
3	**[Check peephole]** If I wasn't expecting visitors and someone knocked on my house/apartment door, I'd check out a window or through a peephole to see who it was before opening the door.	**[Check permissions]** Before installing an app on my phone, I check what the app requests in terms of permissions (e.g., access to my contacts/data, camera, GPS location etc.).
4	**[Protect valuables]** If I lived in my own home/apartment, I'd take measures to hide or protect sensitive information & valuables (e.g. hiding place, safe, firebox)	**[Back up data]** I usually back up data that is important to me at least weekly (e.g., photos, documents)
6	**[Avoid eavesdroppers]** When talking on my cell phone about personal things or sensitive information (e.g. banking), I move to an area away from other people.	**[Check site secure]** When buying things or providing personal information on-line (banking info, SS#), I always check the site is secure (e.g., looking for a padlock icon or "https" in the web address).
7	**[Shred Sensitive Information]** I destroy old papers/materials with personal info prior to throwing them out (e.g., shred or burn old checks, receipts, credit cards, etc.).	**[Postpone Updates]** When my computer tells me a new update is available, I usually procrastinate and don't install it right away.
8		**[Lock cell phone: yes/no]** My phone is set to lock itself when inactive & requires me to type in a passcode/ word or show my face or swipe my fingerprint to use it

TABLE 7.2
Results for Propensity to Trust Computers Scale (from Strongly Disagree = –10 to Strongly Agree = 10)

Computer Trust	Mean		Correlation Coefficients (*r*)					
Item	M	SD	1.	2.	3.	4.	5.	6.
1. 1. I usually trust computers/computer programs until there is a reason not to	4.9	3.9	1					
2. *For the most part, I distrust computers/computer programs*	-4.2	4.3	-.322	1				
3. 3. In general, I would rely on computers/computer programs to assist me	5.2	2.8	.513	-.383	1			
4. My tendency to trust computers/computer programs is high	4.4	3.6	.642	-.473	.758	1		
5. It is easy for me to trust computers/computer programs to do their job	4.9	3.2	.538	-.421	.744	.817	1	
6. I am likely to trust a computer/program even when I have little knowledge about it	2.7	4.4	.366	-.132	.337	.502	.501	1
Mean (item 2 reverse coded)	**4.4**	2.8	**.741**	**-.620**	**.785**	**.908**	**.863**	**.645**

Note: Bold *r* values are significant at the 0.01 level; italics at the 0.05 level, and plain text is *n.s.*

TABLE 7.3
Results for the Computer Expertise Scale (from Strongly Disagree = –10 to Strongly Agree = 10)

Computer Expertise	M	Correlation Coefficients (r)							
Item	[SD]	1.	2.	3.	4.	5.	6.	7.	8.
1. *I don't have much computer expertise. I want it to work and usually don't understand what's going on if there's a problem*	0.9 [5.0]	1							
2. I have a strong understanding of how anti-virus software works	-2.1 [4.6]	-.558	1						
3. I have a strong understanding of phishing & the risk it poses to my personal/professional life	1.6 [4.9]	-.259	.336	1					
4. I have a strong understanding of malware & the risk it poses to my personal & professional life	0.4 [4.8]	-.303	.360	.808	1				
5. I'm proficient at handling cyber-attacks such as phishing or malware	-1.2 [4.7]	-.275	.478	.606	.631	1			
6. I'm proficient at troubleshooting computer SOFTWARE problems	-0.9 [4.8]	-.497	.518	.312	.419	.534	1		
7. I'm proficient at troubleshooting computer HARDWARE problems	-1.4 [5.0]	-.460	.562	.341	.388	.597	.741	1	
8. I often help other people with their computer problems	-0.7 [5.2]	-.546	.510	.346	.439	.482	.666	.576	1
Mean (item 1 reverse coded)	-0.7 [3.6]	-.662	.725	.675	.732	.774	.793	.789	.776

TABLE 7.4
Summary of Security Attitudes and Behaviors Scores Including Percentage of Participants Who Reported Agreement, Descriptives, Comparisons, and Correlations

Item [Physical/Cyber]	%subjects > 0 (%subjects > 5)		Mean [SD]		T-test	Pearson Correlation	
	Phys	Cyb	Phys	Cyb	T	R	P
1. Lock door / Lock computer	91% (71%)	59% (32%)	5.68 [3.99]	1.65 [5.02]	7.77	.153	.083
2. Close curtains / Cover webcam	84% (58%)	32% (20%)	4.59 [4.03]	-2.62 [6.73]	10.81	.067	.449
3. Check peephole / Check permissions	88% (60%)	50% (18%)	5.05 [4.15]	-0.88 [5.79]	9.65	.029	.742
4. Protect Valuables / Back up Data	93% (66%)	45% (20%)	5.67 [3.43]	-0.88 [5.70]	11.57	.068	.442
5. *Hide Spare Key / Re-use passwords*	58% (36%)	91% (63%)	1.21 [5.44]	5.25 [4.12]	-7.12	.103	.245
6. Avoid eavesdroppers / Check site secure	88% (63%)	67% (42%)	5.15 [3.61]	2.49 [5.22]	5.40	**.228**	**.009**
7. Shred sensitive info / *Postpone updates*	59% (32%)	79% (47%)	2.71 [5.11]	3.57 [4.98]	N/A	N/A	N/A
Mean (5 items and cyber 7 item reverse coded)			3.95 [2.49]	-1.30 [2.90]	**17.72**	**.225**	**.010**

Note: Responses were on a slider scale ranging from strongly disagree (-10) to strongly agree (10). T-tests and correlations were two-tailed. For all t values, $p < .001$.

comparisons of agreement/adherence for roughly corresponding items across the subscales; and (iv) correlations across participants on roughly corresponding items across the subscales.

The majority of participants reported some level of agreement/adherence to all good physical security practices except one – specifically, the majority (58%) agreed with the potentially risky practice of hiding a spare key near their home entrance. In contrast, 50% or fewer of the participants reported any agreement/adherence to four of the cyber security practices including: covering the webcam, checking app permissions, weekly data backups, and avoiding password reuse (e.g., 91% reported re-using passwords across multiple apps/sites). T-test comparisons of the mean responses for all corresponding pairs confirmed there was stronger agreement with or adherence to physical security practices than their cybersecurity counterparts. Accordingly, as shown in the last row of the table, in the aggregate, physical security adherence/agreement was higher than cybersecurity adherence/agreement. For the cybersecurity practices the mean value was even slightly negative (M=-1.3), indicating not just a weaker inclination to follow these practices but a slight disinclination to follow them in the aggregate. Thus, the data provided support for the hypothesis and users have a stronger inclination toward physical than cyber security.

It terms of correlations, readers may find it surprising that the correlation was significant for only one corresponding physical-cyber pair listed: avoiding eavesdroppers and checking if [online shopping/banking] sites are secure. Within the physical scale, despite both relating to protecting sensitive information, the tendencies to shred sensitive documents (item 7) and avoid eavesdroppers when discussing sensitive information (item 6) were not significantly correlated (r=.097, p=.274). Furthermore, the means and percentages of participants in agreement indicate that participants are more likely to take steps to avoid eavesdroppers than shred sensitive information. This difference may reflect the fact that potential eavesdroppers are often proximate and concrete threat cues in the moment, whereas individuals seeking to benefit from discarded documents at some future point during trash processing seem anonymous and abstract, and do not present a clear and present danger cue.

For the extra cyber security item in yes/no format about having one's cell phone password or biometrically protected, 91% of participants responded in the affirmative. This sound cell phone security practice in the majority seems at first glance inconsistent with participants' practices regarding locking their computers when leaving them unattended, where only 59% expressed any level of agreement/adherence (response > 0) and less than a third expressed somewhat strong agreement/adherence (response > 5). Note, however, that the items are not about the same stage of security. The cell phone question was about whether password or biometric protection had been set up on their device. However, in the case of computers, users were not asked if their computers had password protection set up (though they all presumably did in order to use the university network). Instead, they were asked whether they (proactively) locked their computer before leaving it unattended. Both cell phones and computers typically are set up to automatically self-lock after some interval of inactivity. However, security risks may be greater for computers for several reasons including but not limited to fact that the required inactivity interval for autolock tends to be longer and one is less likely to leave a cell phone unattended because it is so portable. The item about computer locking better captures potentially risky behavior – that many participants do not make a deliberate point to manually lock or logout of their computers.

7.6   RELATIONS ACROSS SCALES

After any necessary reverse coding, a mean score was computed for each participant to represent their aggregate position for each scale: i) propensity to trust computers; ii) physical security attitudes and behaviors; and iii) cybersecurity attitudes and behaviors (omitting yes/no item 8); and iv) computer expertise. These values were used to compute overall inter-scale correlations. As might be expected, good cybersecurity practices were correlated negativity with a user's trust in computers ($r = -.340$, $p <.001$) and positively with their computer expertise ($r=.383$, $p<.001$), and their physical security practices ($r=.225$, $p=.010$). However, note that the physical security score was the weakest of these three predictors of cybersecurity attitudes/ behaviors, and the relationship was only significant in the aggregate. Recall from Table 7.2, that pairwise, the physical and cyber security item counterparts were typically not significantly correlated.

High physical security agreement/adherence did not predict decreased trust in computers ($r=.101$, $p=.251$), nor did it predict increased computer expertise ($r=.012$, $p=.894$), contrary to the expectation that having a (physical) security mindset might make one more wary of computers, and might serve as a motivator to gain cybersecurity expertise. Finally, although increased computer expertise might be expected to make one more aware of the cyber security risks, it was not significantly [negatively] related to the propensity to trust computers ($r=-.095$, $p=.280$).

7.7   DISCUSSION

Even in our current information age, the data indicate that levels of agreement with and/or adherence to physical security practices (e.g., lock door, close curtains, check peephole, etc.) are much stronger than agreement/adherence levels for roughly analogous cybersecurity practices (e.g., lock computer, cover webcam, check app permissions, etc.). For the cybersecurity practices the mean agreement value was even slightly negative, indicating not just a weaker inclination to follow these practices but a slight disinclination to follow them. These findings raise two important questions, which are discussed, in turn, in the following subsections. The first question is: Why is there a stronger disposition toward physical security than cyber security? The second question is: Can these results and/or the underlying factors producing them inform an effective approach for improving cybersecurity practices?

7.7.1   WHY IS THERE A STRONGER DISPOSITION TOWARD PHYSICAL VERSUS CYBER SECURITY?

Before focusing on the explanatory factors favored by author, based on the view that they best lend themselves to informing the design of novel potential interventions and interfaces to promote cybersecurity practices, the discussion first explored some other possible factors that could contribute to this pattern of prioritization of physical security over cybersecurity.

One might argue that young adults may be relatively unconcerned about some aspects of security in general because they have yet to accrue many big secrets, or digitally accessible assets and intellectual property. However, this might be expected to produce laxity in some aspects of physical as well as cyber security, because they also would have yet to accrue many valuable physical artifacts. Nonetheless they report a strong tendency to protect and hide the [physical] valuables they have (M= 5.67, 93% agreement). Furthermore, among cybersecurity items, there is relatively high adherence to protecting financial assets by checking that sites are secure (M= 2.49, 67% agreement) relative to adherence to other cyber security practices. Thus, low net worth may be an insufficient explanation of the data patterns and, even if it were a factor, it is not a feasible approach to improve users' cybersecurity practices by providing them with more [digital] assets to protect!

Another possible factor could be cultural - based on the type and amount of information that young adults now share on social media. Some research (e.g., Steijn, Schouten & Vedder, 2016) has explored whether participants who have grown up in the information age might have a reduced concern about privacy or might characterize privacy differently than older adults. If so, the data in the current study suggest that any reduced concern about security and privacy seems limited to cyber domain security practices. Note that the majority of participants still agreed with/ adhered to practices like locked doors, closed curtains, and avoiding eavesdroppers. While young adults may be comfortable voluntarily sharing some information about themselves on social media, they clearly still value many aspects of privacy, and they may not realize the degree to which computer and social media use can provide others with access to information (and assets) that the user would not want to voluntarily share.

This last point highlights possible education and awareness factors. Poor cybersecurity practices are often attributed to deficits in risk awareness, knowledge of best practices, and/or a lack of motivation to follow them (Frey, 2018). Such factors suggest that an effective approach to improve cybersecurity practices is via campaigns to increase knowledge of best practices and awareness of risks, and/or introduce incentive or penalty systems for compliance (e.g., Goel, Williams, Huang & Warkentin, 2021). However, cybersecurity awareness programs often fail to change behavior (Nagyfejeo & Von Solms, 2020), and extrinsic (e.g., financial) incentives for compliance can be costly and may be less sustainable than intrinsic motivation. Note that people do not need to be paid to close their curtains nor lock their doors.

As foreshadowed in the introduction, the author instead advocates shifting focus to other possible explanatory factors. In particular, physical threats are often more visually apparent - or at least more readily visualizable - to support salience and Situational Awareness (see also Pyke, Bouchelle & Uzhca, 2023; Pyke, Feltner & Ness, 2023). Additionally, physical threats tap into to fundamental needs and motivations (i.e., the Safety Tier in Maslow's hierarchy), and may be bootstrapped or amplified by some visceral evolutionary associations. Beyond being plausible, as discussed in the next section, these explanatory factors are primarily important because they afford opportunities for the design of interfaces or interventions to improve cybersecurity practices.

7.7.2 IMPLICATIONS FOR THE DESIGN OF CYBERSECURITY INTERFACES/ INTERVENTIONS

Cybersecurity threats like remote hackers, backdoors and computer viruses are not directly observed by users, and consequently might be not only out of sight but also out of mind. A stronger impetus toward physical security practices might be due, in part, to physical threats being more visible (or visualizable; e.g., Pyke, Bouchelle & Uzhca, 2023). Just as interfaces are being developed to support military decision makers by providing real-time visible cues to represent cyber threats (e.g., Llopis et al., 2018), it would be possible to provide regular computer users with an interface designed to provide visible cues about current cybersecurity threats to trigger cybersecurity behaviors. In the current study, the cybersecurity practice with the greatest self-reported adherence was checking that a website was secure, and this cybersecurity threat is one that is represented by the presence or absence of a visible lock icon cue in the interface. Even so, only 67% of participants expressed any level of agreement with/adherence to checking site security, and only 46% reported moderate to strong adherence.

Thus, while a good start, providing visible cues for cybersecurity threats may be insufficient to strongly influence cyber security practices. To be particularly effective, cues may not only need to be visible, but they might also need to provoke a visceral reaction like some physical threat cues do (e.g., when we notice a stranger staring at or apparently following us). As discussed in the introduction, some FPS video game interfaces (e.g., Call of Duty, Battlefield, Escape from Tarkov) simulate cues associated with physical threat or injury such as displaying your (avatar's) blood spatter, or a cracked screen effect to represent damage to your (avatar's virtual) face shield. Similarly, a cybersecurity interface could display cues for cybersecurity threats that simulate or resemble physical threat cues. For example, if a security update is required, the screen could display crack with an eyeball looking in that widens over time the longer the update is postponed. Audio cues could also be used to signal cybersecurity threats, like the sound of someone jiggling the handle on a door attempting to get in. Education and training may be insufficient to activate the visceral reactions and motivations that cues of mimicking those of physical threats might be able to produce.

7.8 CONCLUSIONS

The results suggest that poor cyber security practices in young adults do not reflect a general laxity about security that also extends equally to physical security practices. Rather, reported adherence to physical security practices is considerably higher than adherence to roughly analogous cybersecurity practices. Consequently, it may be possible to improve cybersecurity practices by harnessing the cues and mechanisms that drive physical security practices. Physical threats tend to be more visible than cyber security threats (e.g., remote hackers), and physical threat cues also tend to provoke more motivated, visceral responses. An interface that represents cyber security threat cues in a manner that resembles physical threat cues may hold promise for improving cybersecurity practices. For example, just as a person is likely to shut their curtains if they notice someone peering in, a person may be more likely to cover or disable their webcam if their screen displays a (simulated) hacker looking back at them.

Disclaimer: The views and opinions expressed herein are those of the author own and do not necessarily state or reflect those of the US Army, Department of Defense or the United States Government.

Acknowledgements: The author would like to thank Trey Martin, Elizabeth Rodriguea and Alanna Severt for their assistance with data collection, and adapting the Propensity to Trust Technology Scale to focus on computers/programs more specifically.

REFERENCES

El-Bably, A. Y. (2021). Overview of the impact of human error on cybersecurity based on ISO/IEC 27001 information security management. *Journal of Information Security and Cybercrimes Research, 4*(1), 95–102.

Endsley, M. R. (1988, October). Design and evaluation for situation awareness enhancement. In Proceedings of the Human Factors Society annual meeting (Vol. 32, No. 2, pp. 97–101). Los Angeles, CA: Sage Publications.

Endsley, M. R. (1995). Measurement of situation awareness in dynamic systems. *Human factors, 37*(1), 65–84.

Fischer, R. J., Halibozek, E. P., & Walters, D. A. (2013). Introduction to Security, Waltham, MA.

Frey, S. (2018). How to Eliminate the Prevailing Ignorance and Complacency Around Cybersecurity. *Cybersecurity Best Practices: Lösungen zur Erhöhung der Cyberresilienz für Unternehmen und Behörden*, 1–10.

Goel, S., Williams, K. J., Huang, J., & Warkentin, M. (2021). Can financial incentives help with the struggle for security policy compliance?. *Information & management, 58*(4), 103447.

Kellerman, A. (2014). The satisfaction of human needs in physical and virtual spaces. *The Professional Geographer, 66*(4), 538–546.

Lee, S. Z. (2020). A basic principle of physical security and its link to cybersecurity. *International Journal of Cyber Criminology, 14*(1), 203–219.

Llopis, S., Hingant, J., Pérez, I., Esteve, M., Carvajal, F., Mees, W., & Debatty, T. (2018, May). A comparative analysis of visualisation techniques to achieve cyber situational awareness in the military. In *2018 International Conference on Military Communications and Information Systems (ICMCIS)* (pp. 1–7). IEEE.

Merritt, S. M., Heimbaugh, H., LaChapell, J., & Lee, D. (2013). I trust it, but I don't know why: Effects of implicit attitudes toward automation on trust in an automated system. *Human factors, 55*(3), 520–534.

Nagyfejeo, E., & Von Solms, B. (2020). Why do national cybersecurity awareness programmes often fail. *International Journal of Information Security and Cybercrime, 9*(2), 18–27.

Poston, B. (2009). Maslow's hierarchy of needs. *The surgical technologist, 41*(8), 347–353.

Pyke, A., Bouchelle, R., Uzhca, D. (2023). Out of Sight but Still In Mind: Making 'Invisible' Cyber Threats More Salient Via Concrete Analogies. In: Abbas Moallem (eds) Human Factors in Cybersecurity. *AHFE Open Access*, vol 91. AHFE International, USA.

Pyke, A., Feltner, D. & Ness, J. (2023). What Types of Tactical Vulnerabilities do Future Officers Most Anticipate: Are Cyber as well as Non-Cyber Threats on their Radar? *Cyber Defense Review*, 8(1), 103–117.

Richards, C. (2020). Boyd's OODA loop. *NECESSE (Royal Norwegian Naval Academy Monographic Series)*, 5(1), 142–165.

Smith, D., Leonis, T., & Anandavalli, S. (2021). Belonging and loneliness in cyberspace: impacts of social media on adolescents' well-being. *Australian Journal of Psychology*, 73(1), 12–23.

Steijn, W. M., Schouten, A. P., & Vedder, A. H. (2016). Why concern regarding privacy differs: The influence of age and (non-) participation on Facebook. *Cyberpsychology: Journal of Psychosocial Research on Cyberspace*, 10(1).

Section IV

Perspectives on Linguistic
Approaches in Cybersecurity

8 Using Language Translation Software to Detect and Classify Cyberattacks

Wayne Patterson, Carlos Azzoni, Diana Florea, Kaido Kikkas, Birgy Lorenz, Rosangela Malachias, Leonid Vagulin, and William Emmanuel S. Yu

8.1 THE PROBLEM

With the increasing worldwide access to the Internet, the possibilities of cyberattack have increased exponentially. The purpose of this paper is to expand and improve on a technique that can assist less skilled computer users in defending their environment against potential cyberattacks.

DOI: 10.1201/9781003599142-12

The perspective of this study is that there can be a distinction in the level of ability for either a defender or attacker in a cyberattack.

The best way of describing this distinction is to describe the interaction between attacker and defender using a "game theory" model.

Let us assume we can divide the universe of attackers and defenders reduced to two categories of each type. For the purpose of this model, we will divide all potential attackers into the "expert" and the "naïve" category.

You can do the same with the defenders. We will give a few examples as to how this categorization should be made:

The "players" in this game can be considered, on one side, the Cyberattackers and on the other the Cyberdefenders. A 2x2 model of the players follows:

		Cyberattackers	
		Naïve	Expert
Cyberdefenders	Naïve	Hackers, …	Government agencies, eg NSA (US), Mossad (Israel), KGB (Russia)
	Expert	Gov. agencies, Corporations, Computer companies	

Using the game theory approach, we assign a cost and benefit in the game. For example, if an Expert Cyberdefender is attacked by an Expert Cyberattacker, the cost to each might be, say, $100,000. If each of the attacker and defender is Naïve, the value might be minimal, say $10.

This implies that the Expert Cyberattacker is unlikely to choose to attack the Naïve Cyberdefender because the cost of launching the attack is likely to require more resources (dollars) than the expected reward.

		Cyberattackers	
		Naïve	Expert
Cyberdefenders	Naïve	$10, -$10	0,0
	Expert	0,0	$100K, -$100K

8.2   DETECTING INTERNATIONAL PHISHING OR RANSOMWARE ATTACKS

Over the past decade, the number of cyberattacks such as ransomware, phishing, and other forms of malware have increased significantly. The ability to launch such devastating attacks is no longer limited to highly structured organizations including government agencies whose missions may well include cyberattacks. The focus of our study is on threats to an individual not from such organizations, but rather less organized cybercriminal groups with limited resources. The Internet provides ample opportunities for such criminal organizations to launch cyberattacks at minimal cost.

One tool for such lower-level criminal organizations is Google Translate (GT) needed to launch a cyberattack on a user in a relatively advantaged country such as the United States, United Kingdom, or Canada. It or has been observed that many such attacks may originate in a lesser developed country (LDC), where the local language is a language not common to persons in target countries, for example English.

It is a reasonable assumption that informal cyberattackers may not have a command of English and to use English for an attack online they may require a mechanism, such as the no-cost GT.

The purpose of this paper is to propose a series of techniques that can provide an approach to determining if a malicious cyberattack, such as ransomware or phishing, can be determined as to the attack language or country of origin. Many attacks of the types described above, in communicating with the target, must transmit text in some format. For example, in a ransomware attack, the target must be told that certain steps are necessary for the ransomware to be paid. Similarly, a phishing attack will also normally try to trap the target into a response that will provide a benefit to the attacker.

In most cases, cyberattacks of this nature must provide some information as text in a human language, in order to describe to the potential victim the actions that the attacker expects to be carried out. Since there are literally hundreds of natural languages throughout the world, it is entirely possible and indeed likely that the cyberattacker will need to provide some text in the language of the persons being targeted. Given that the potential attacker can come from virtually any corner of the world, it will often be the case that the cyberattackers command of his or her language will not include the language of the target. Thus, it will be necessary that any text information will need to be first translated from the human language of the attacker to a human language presumed to be understood by the victim.

With the existence of the Internet, we must assume that any cyberattack may originate in any part of the world, and thus also that the attacker needs to have a mechanism for his or her expected target to be able to read and understand the attack text in their own language, in order for the attacker to present the challenge to the potential victim, and thus potentially reap the reward. Given the large number of human languages in use throughout the world, if we wish to narrow down the potential source over a attack, it would be useful if we could examine the attack text and try to determine if had been translated between two human languages. In such a case, a fairly definitive conclusion could isolate the potential source of the attack.

8.3 ABA TRANSLATION

In order to develop an approach to determine if a given body of text had been originally translated between human languages, we developed a technique called ABA translation. Using English as a base for our various analyses, we developed a body of English text from two categories: classical quotations (L) that are well known in English, and that demonstrate grammatically correct use of the English language;

and a category of popular expressions from more colloquial English language, in particular popular film (F); thus in this text we will refer to "LF".

8.4 LEVENSHTEIN DISTANCE

In order to determine a metric for the quality of the translation based on the text samples, we use a well-known approach called the Levenshtein Distance to determine the accuracy of the ABA translation. Given this approach we can determine a metric with the assumptions above, for the translation of text between any of the given language pairs. For our purposes, we use a version of the standard definition that we designate as Modified Levenshtein Distance (MLD). In this case, we break up the computation of this distance for substrings of all the strings being compared. Thus the opening few characters of the strings being compared will not propagate through the entire set of strings. Here are a few examples to demonstrate computing MLD, using three of the languages used in this current study.

Russian ([1] Russian --- [2] Original English --- [3] Translation Back to English

```
[1] Ваш IP-адрес использовался для посещения веб-сайтов,
содержащих порнографию, зоофилию и жестокое обращение с
детьми.
```

```
[2]Your IP address was         used to visit websites
containing pornography, zoophilia , and child abuse.
```

```
[3]Your IP address has been used to visit websites that
contain     pornography, bestiality, and child abuse.
```

```
/   7/           /    4/    /   3/          /       10/
                 MLD = 7 + 4 + 3 + 10 = 24
```

Romanian ([1] Romanian - [2] Original English - [3] Translation Back to English

```
[1] Spune-le să meargă acolo cu tot ce au şi să câştige
doar unul pentru Gipper.
```

```
[2] Tell   'em to go out there with all      they got   and
win just one for the Gipper.
```

```
[3] Tell them to go      there with everything they have
and win just one for     Gipper.
```

```
/   2/      /      3/    /      10 /   /    4/    /     3/
                 MLD = 2   + 3 + 10 + 4 + 3 = 22
```

Portuguese (Brazilian) ([1] Portuguese - [2] Original English - [3] Translation Back to English

[1] Se você acha que pode, ou que não pode, geralmente está certo.

[2] Whether you think that you can, or that you can't, you are usually right.

[3] If   you think   you can, or   you can't, you' re usually right.

```
/     7/       /      4/        /      4 /      /     1/
                          MLD = 7   + 4 + 4 + 1 = 16
```

In previous work, a number of authors have attempted to develop an index to measure the efficiency or what might be called an ABA translation. This involves beginning with a test document in language A (see Table 8.1), then GT to translate into language B, then back again into A. The resulting original text is then compared to the transformation by using a modified Levenshtein distance computation for the A versions. The first ten messages are drawn from a list of messages arising from actual attacks; the other 20 are from a list originally used in (Patterson 2021).

The paper analyzes the process of determining an index to detect if a text has been translated from an original language and location, assuming the attack document has been written in one language and translated using GT into the language of the person attacked. The steps involved in this analysis include:

a. Consistency: in order to determine consistency in the use of the ABA/GT process, the primary selection of test is compared with random samples from the test media;
b. Expanded selection of languages for translation: prior work has established use of the technique for 12 language pairs (Patterson, 2021), (Florea and Patterson, 2021). The current work extends analysis to a wider set of languages, including those reported as having the highest levels of cyberattacks. This component is also addressed in greater detail in (Blackstone and Patterson, 2022).
c. Back translation of selected languages: used to extend the quality of those translations are made.
d. New language pairs are considered: by analyzing the countries and indigenous languages of the countries paired with the highest levels of cyberattack and the highest levels of cyberdefense, additional language pairs are added to this analysis;
e. Comparison to prior results: results found in this paper are used for a proposed network for all language pairs considered in this analysis.

The end product is a metric giving a probability of determining the original source language of the cyberattack as compared to the translation to the victim's language, with the expectation that this will allow for an increased likelihood of being able to identify the attackers.

TABLE 8.1
List of Attack Messages Used

Attack Messages (Cyber)

1	Your important files are encrypted.
2	Maybe you are busy looking for a way to recover your files, but do not waste your time.
3	You can decrypt some of your files for free.
4	But if you want to decrypt all of your files, you need to pay.
5	Payment is accepted in Bitcoin only.
6	Encryption was produced using a unique public key.
7	Any attempt to remove or damage this software will lead to the immediate destruction of the private key by server.
8	Your IP address was used to visit websites containing pornography, zoophilia, and child abuse.
9	This computer lock is aimed to stop your illegal activity.
10	To unlock the computer you are obliged to pay a fine.

Attack Messages (Literature and Film)

11	I'm as mad as hell, and I'm not going to take this anymore!i
12	When a person suffers from delirium, we speak of madness. when many people are delirious, we talk about religion.
13	Of all the gin joints in all the towns in all the world, she walks into mine.
14	Open the pod bay doors, please, HAL.
15	Mrs. Robinson, you're trying to seduce me. Aren't you?
16	Keep your friends close, but your enemies closer.
17	If you build it, he will come.
18	A lie gets halfway around the earth before the truth has a chance to get its pants on.
19	I have always depended on the kindness of strangers.
20	Sex and divinity are closer to each other than either might prefer.
21	Political correctness is despotism with manners.
22	The only way to get rid of a desire is to yield to it.
23	Whether you think that you can, or that you can't, you are usually right.
24	There are no facts, only connotations.
25	I'm living so far beyond my income that we may almost be said to be living apart.
26	People demand freedom of speech to make up for the freedom of conviction which they avoid.
27	Tell 'em to go out there with all they got and win just one for the Gipper.
28	Round up the usual suspects.
29	Love means never having to say you're sorry.
30	The greatest glory in living lies not in never falling, but in rising every time we fall.

8.5 REFERENCE TO PRIOR STUDY

Rather than attempting to detect any text that may have been translated between any pair of languages, we have an interesting source identifying those countries that have received the greatest frequency of cyberattacks. This research (Bruce, Lusthaus et al, 2024), called the World Cybercrime Index (WCI), has identified the ten countries with the greatest level of cyberattacks, and this is summarized in the following Table 8.2.

TABLE 8.2
Comparison of Test Countries (Including WCI)

Country	Rank	WCI	Languages	Computer %	Population (Millions)
Russia	1	58.4	Russian 85.7	89.5	140.8 M
Ukraine	2	36.4	Ukrainian 67.5, Russian 29.6	65.6	35.7 M
China	3	27.9	Mandarin	77.3	1,416.0 M
Romania	6	14.8	Romanian 91.6	64.9	18.1 M
North Korea	7	10.6	Korean	---	26.3 M
Brazil	9	8.9	Portuguese	77.1	220.1 M

There have been several papers written on topics similar to this one. One notes, for example, (Patterson, 2021), (Florea and Patterson, 2021), (Blackstone and Patterson, 2022). However, research by Bruce and Lusthaus (Bruce, Lusthaus et al, 2024) has provided a methodology for a more detailed analysis of sources like countries and nationalities --- and therefore, to a higher likelihood, the choice of human language familiar to the cyber attacker. The aforementioned authors indicated in their paper "This data was carried out as a partnership between the Department of Sociology, University of Oxford and UNSW Canberra Cyber". Table 8.2 shows the comparison of test countries.

8.6 SELECTION OF COUNTRY IN CYBERATTACKS

If we assume that the analysis in the previous section is valid, the question arises as to how a user, presuming that an external attack may have taken place in their system, may use the results of this analysis to allow the person attacked to develop a response to this attack.

The first step for this person might be to analyze the attack message (that might be a phishing attack, ransomware, or another) to see how the results may assist in at least determining the likely origin of the attack by country.

First, copy the suspect text to a new workspace. Then, translate the suspect text, using GT and the ABA process, to each of the six suspect languages.

Now, calculate each of the ABA scores, and place them on the graph in Table 8.3 below. There is a principle called "maximum likelihood" that holds that if we have distances tabulated for each of the suspect languages, the greatest likelihood is the one that will be the closest distance to the known values. As an example, note the conclusion from the following Six Country Comparisons.

8.7 INTRODUCING TECHNIQUE GT

The question can now be raised, how can the nonspecialist user make use of the techniques described in this paper, in order to determine if he or can detect if she or he can detect the source of the Cyberattack from one of the indicated countries.

If so, the following suggestions for the user who has access to the Cyberattack and who may have read this paper.

1. If the user suspects that there is a Cyberattack, he or she may attempt to determine if the attack is from one of the countries described in this paper. The first step should be to make a copy of what the user may suspect to be the attack text sentences.
2. Then using the translation as the example used in this paper, the user should use the ABA process to the received copy, then to the suspect language, then back to the attack language.
3. For each of the six suspected attack countries, the Google Translate algorithm should be computed.
4. Then back to the suspect text, who with the lowest score should be initially considered the attacker.

8.8　RESULTS

We will assume that the readers of this paper can be generally considered (relatively) Naïve Cyberdefenders. If an Expert Attacker, the cost to each might be, say, $100,000. If each of the attacker and defender is Naïve, we will also assume that the Naïve Cyberdefender does not have the resources to be able to defend against the Expert Attacker; however, this is not a great risk because the Expert Attacker is unlikely to be interested in attacking the Naïve Cyberdefender, because the reward is likely to be very small.

Thus, for the focus of this paper, we will assume that we are a Naïve Cyberdefendehr and we need to build a defence against a Naïve Attacker.

We can use the Attack Messages of Table 8.1 to assess the levels of attack for the Cyber data (C) and the Literature and Film data (LF). The result of the GT and ABA exercises is described in Graph 1 (Six Country Comparisons) below.

This table is constructed from the actual values in Table 8.3, then the data is normalized so that each C and LF has modified in the range [0,1].

Country Abbreviations:

Brazil	BR	China	CH	North Korea	NK
Romania	RO	Russia	RU	Ukraine	UK

TABLE 8.3
Comparison of Test Countries (Including WCI)

Country	LF	C	LF Normalized	C Normalized
Russia	338	82	82.5	17.7
Ukraine	264	82	26.7	17.7
China	440	131	1.0	37.3
Romania	337	84	.57	18.5
North Korea	200	287	0.0	100.0
Brazil	201	38	0.4	0

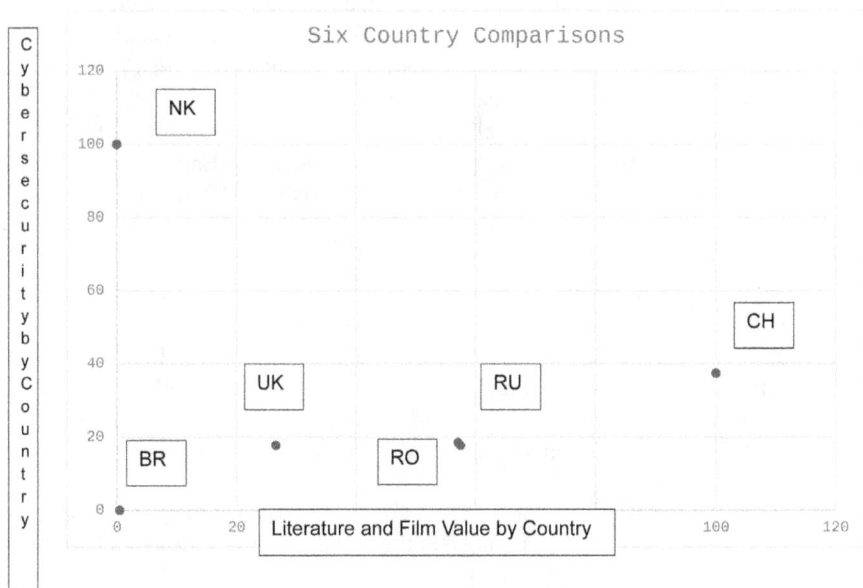

Six Country Comparisons

8.9 COMMENTARY ON GT TRANSLATION QUALITY FOR SELECTED LANGUAGES

Several of the co-authors of this paper have contributed analyses of the quality of the translations provided by GT. The original texts quoted in each language can be found in Table 8.1. Their comments follow:

(a) ABA = English – Russian – English

Comment by Kaido Kikkas:

Did a first reading and spotted an interesting error in "Open the pod bay doors, please, HAL" at once the GT apparently considered the "pod bay" a location, leaving it untranslated and writing it in Latin script. I find it interesting that Russian language has been shifting from writing foreign names (and sometimes terms) in Cyrillic the way it is pronounced (e.g. "Sherlok Holms") to writing it verbatim in Latin script. However, a number of foreign terms are also written in Cyrillic (e.g. "website" => "veb-sajt").

As my first impression, it seems that

* the backtranslation (Russian to English) is generally better than the first translation from English to Russian (probably no wonder, as English is likely the predominant language for Google). But there were exceptions - e.g. in 16, the Russian translation has correctly retained the original "but", the backtranslation has turned it into "and". And see the remark on 13 below.

* some mistakes are subtle (e.g., the very English "there are" is left untranslated, which somewhat changes the meaning (the English original has the connotation of "no facts exist", while the Russian translation is more "in this case, there are no..."). However, some are more direct, e.g. in 17 and 30 the meaning of the whole sentence haschanged.
* the most messy one is in "Of all the gin joints in all the towns in all the world, she walks into mine." - at first "gin" turns into "jeans" in English-Russian, and then "jeans" turn into "denim" in Russian-English (the latter is more subtle, a garment vs a material). Also, here is another Russian-English mistranslation - "she walks into mine" is correctly translated in Russian, mistranslated in the English backtranslation ("she is mine").
* the translations in the first part (attack messages) are rather good (only some small emphasis changes or using (near-) synonyms). The GT seems to have more problems with the literature and film phrases - I'd explain it as attack messages having a more direct (and somewhat narrowly technical) nature, while there is more creativity and context in the literary language.

While the Russian translations made by GT are correct as such, they are worded so that in a real attack message they do not 'hit the nerve', i.e. the victims will not react as the attacker wants.

Note: I have seen similar attempts on myself in Estonian language which also betray the writer in certain ways.

Besides, in large languages there are big differences in style between e.g. an academic and a simple laborer - what works on one, won't work on the other (both ways).

Comment by Leonid Vagulin:

[Examples given were in part from a list of malware expressions; the others from English film and literature references.]

Malware part: Since the malware part is somewhat coherent the translation adjustment provided are in a scope of "encryption malware information section". Some words were omitted since they carried the target away from the intimidation/persuasion. Adjusted translation was done assuming that malware text is taken from a single source and context. I tried to make the malware translation universal by exploiting human simple yet descriptive phrases.

Film and Literature part: This translation was tricky, since the context on some inputs was not clear and required googling the source, usages.

Comment by Birgy Lorenz:
My question and suggestion are to do the same with AI tools, meaning to do comparison and discuss the changes from Google to AI as well.

As in Estonia no one really uses to translate Google anymore, now AI is used to come up with these scams. Or to do whatever.

This would also raise the interests and modernize the chapter.☺

(b) ABA = English – Chinese – English

Comment by William Yu:
Unfortunately, I might have a Chinese last name and am technically ethnic Chinese. I do not know the language well outside of a few classes when I was in grade school. So I am not certain I would be of much help.

Some comments on the chapter:

- Mainland China (PRC) generally uses simplified Chinese characters. I also noticed that a lot of tech workers in the field use both english and simplified characters.
- I believe Taiwanese (ROC) used to use traditional Chinese characters more. So there might be a small correlation you can take advantage
- Both simplified and traditional Chinese characters generally have a one to one mapping. So it really is not an issue.
- Something to consider is that threat actors may actually be vectoring attacks to countries like China and Russia but are actually from other countries. People tend to question the source less if they know it is from these two countries. I also recall a feature on North Korea's cybercrime industry being based and/or routed via China. So if the source is Dandong, China that might actually be North Korea.

(c) ABA = English – Romanian – English

Comment by Diana Florea:
Using the ABA technique, for the most part, the translations from English do somewhat retain the proper meaning in Romanian and backwards.

However, in a number of cases, the Google Translate (GT) Romanian renditions would appear to a native Romanian speaker: awkward "I'm as mad as hell, and I'm not going to take this anymore!", "A lie gets halfway around the earth before the truth has a chance to get its pants on." (11,18); incomplete (18,26); sound totally non-native (12,23, 25); non-sensical as in losing the original meaning (13) or presuppose inside cultural knowledge of movie/play/script (14). Furthermore, in all these cases, the GT tool provides formal translations as it typically aims to provide a general translation that is correct and understandable, rather than focusing on specific nuances like formality levels. More to the point, Romanian being a language with more distinct differences between formal and informal language that GT may try to capture — for example the so-called *pronume de politeţe* 'pronouns of politeness' which allow to express different degrees of politeness, formality or familiarity — it calls for a translation environment that must be determinedly clear when politeness hence hierarchies are intended, see for reference the Romanian GT table examples "You can decrypt some of your files for free." (3, 8). In conclusion, albeit Romanian variants seem to adhere to the conventions of everyday speech, the ABA technique yields critical information on the language of origin used by a potential cyber attacker.

(d) ABA = English – (Brazilian) Portuguese – English

Comment by Carlos Azzoni:

Ten problems in the translation can be identified. Here are 5 in detail.

1. The use of "gin joints" is suggested by the examples as "De todas as juntas de gin em todas as cidades do mundo, ela entra na minha", but more properly "De todos os bares em todas as cidades do mundo, ela entra na minha."

2. In English, "If you build it, he will come." The program suggests "Se você construí-lo, ele virá", but should be "Se você construí-lo, a pessoa virá." In this sentence, there is a confusion between the "it" and the "he". You build something and some person will come. It is grammatically correct in Portuguese, but it distorts the meaning."

3. "A lie gets halfway around the earth before the truth has a chance to get its pants on." Translates into "Uma mentira fica no meio do caminho antes que a verdade tenha a chance de obter" but should be "Uma mentira dá meia volta ao mundo antes de que a verdade tenha tempo para vestir as calças."

This is probably the biggest mistake in all the sentences. The translation indicates that the lie will fall short of the way before the truth has a chance … .

4. Political correctness is despotism with manners. "O politicamente correto é o despotismo comos costumes" should be "O politicamente correto é o despotismo com bons modos."

Here the translation changed the meaning of the sentence. It substituted despotism with the manners (as for being hard on the use of manners) for despotism with manners (as in despotism with elegance) …

5. "Tell'em to go out there with all they got and win just one for the Gipper" is translated by the software as "Diga para ir lá com tudo o que eles têm eganhar apenas um para o Gipper", better would be "Diga-lhes para darem tudo o que têm eganharem uma pelo Gipper."

The problem is more on the Portuguese / English situations. As in French and in Spanish, we tend to locate the subject of the sentence not in the beginning of the sentence, but in the end (or in the middle). Typically, the translator keeps looking for the subject and usually does not find it. Passive voice is also highly used in Portuguese (As palavras foram trocadas). The words have been changed, instead of We changed the words …).I hope I understood the task. Please let me know if you need anything else.

Comment by Rosangela Malachias:

I do not know anything about cybercrime, but I already thought it would be possible to avoid mistakes as "face recognition" that, here, in Brazil, is a racist method of Police to "arrest Black (innocent) people".

In my work, we, the staff, already saw fraud from white candidates who edit pictures to become themselves "pardos" /mixed in order to ask for quotas.

I think that GT is an important Tool to facilitate connections and communication, but it sometimes do not work very well translating regular Portuguese, slangs and regional expressions.

Let me know if my upset about fraud in the affirmative action system and the denouncement about face recognition can be considered issues in an article. In this case, GT can be a source for English references readings.

REFERENCES

Blackstone, J. and Patterson, W. (2022). Isolating Key Phrases to Identify Ransomware attackers, to appear.

Bruce, M., Lusthaus, J., Kashyap, R., Phair, N., and Varese F (2024). Mapping the Global Geography of Cybercrime with the World Cybercrime Index. *PLoS ONE* 19(4): e0297312. https://doi.org/10.1371/journal.pone.0297312.

Florea, D. and Patterson, W. (2021). "A Linguistic Analysis Metric in Detecting Ransomware Cyber-Attacks". *International Journal of Advanced Computer Science and Applications (IJACSA)*, 12: pp. 517–522. https://doi.org/10.14569/IJACSA.2021.0121158

Patterson, W. (2021). "Detecting Cyberattacks Using Linguistic Analysis", Advances in Human Factors in Robots, Unmanned Systems and Cybersecurity, Matteo Zallio, Carlos Raymundo Ibanez and Jesus Hechavarria Hernandez, eds., *Proceedings of the AHFE 2021 Virtual Conference on Human Factors in Robots, Drones and Unmanned Systems, and Human Factors in Cybersecurity*, July 25–29, 2021, USA, pp. 169–175.

9 Isolating Key Phrases to Identify Ransomware Attackers

Jeremy Blackstone

9.1 INTRODUCTION

Ransomware attacks are a class of cyberattacks that prevent a user from accessing data on their device until the user provides some type of compensation. There are millions of cases of ransomware every year and advancing this field will protect consumers, private companies and government organizations (Simoiu et al., 2019).

9.2 BACKGROUND

9.2.1 THREAT MODEL

We assume that the adversary is a non-native English speaker and that the final ransomware message is in English translated via Google Translate. We assume this because using a language pervasive throughout the internet provides an adversary with more potential targets and a large volume of ransom notes are translated via Google Translate as shown in (Florea and Patterson, 2021).

9.2.2 LEVENSHTEIN DISTANCE AND ABA TRANSLATION

Levenshtein Distance (LD) is a technique used in information theory used to quantify the difference between two string sequences (Levenshtein, 1966). In this study we perform an ABA translation as shown in example 1.

 DOI: 10.1201/9781003599142-13

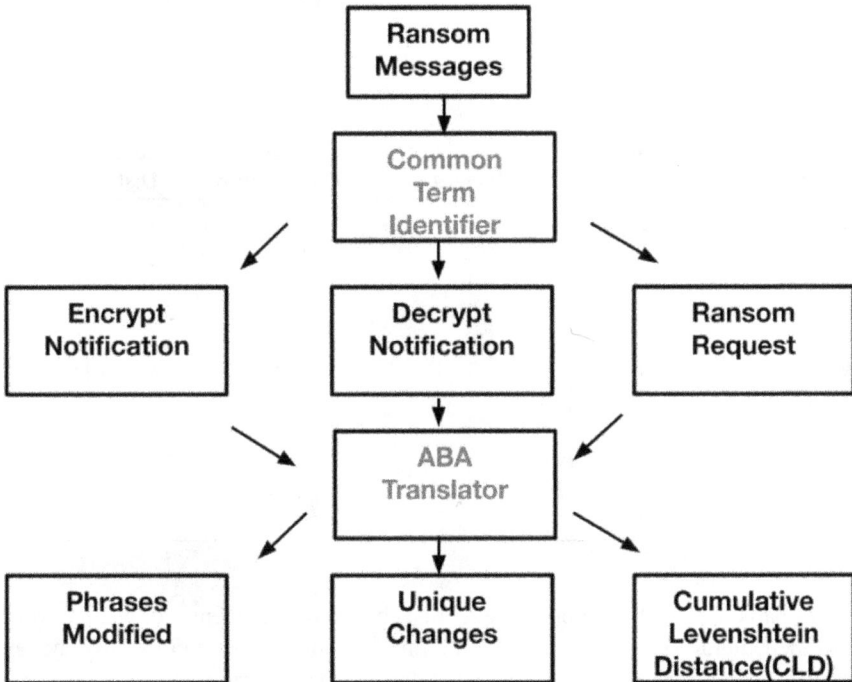

FIGURE 9.1 Workflow for isolating and analyzing key phrases.

Example 1: English-Chinese(Simplified)-English

A: To decrypt your files
B: 解密你的文件
A: To decrypt **[your]** file**[s]**
LD: 'y'+'o'+'u'+'r'+' '+ 's' = 6

9.3 EVALUATION

We conduct our investigation based on the process shown in Figure 9.1. We analyzed the text of ransom messages used for notable ransomware attacks and identified common terms used in all of the messages. We ranked the languages from most to least distinguishable by performing an ABA translation, calculating the LD and determining the number of phrases modified, number of unique changes made, number of unique combinations of changes as shown in Table 9.1.

9.4 CONCLUSIONS

We generated phrases for commonly used in ransom messages in English and performed an ABA translation on all languages included on Google Translate and found that select languages differ from the original message because the languages

TABLE 9.1
Most Easily Distinguishable Languages

Language	# Phrases Modified	# Unique Changes	# Unique Combinations	Cumulative Levenshtein Distance (CLD)
Uyghur	2	2	1	17
Chinese	2	2	1	12
Slovenian	2	1	1	16
Hungarian	2	1	1	10
Korean	2	0	1	9
Kyrgyz	1	1	0	15
Azerbaijani	1	1	0	11
Maori	1	1	0	7
Czech	1	1	0	7
Hawaiian	1	1	0	4
Arabic	1	1	0	2

change parts of speech, add or delete words, change the verb tense, or change words from their plural to singular form. Using this information, we can identify the language of origin for a new message based on whether it prefers or omits the types of grammar we observed in our analysis. While this study focused on ransomware messages, similar analysis could be applied to phishing messages as indicated in (Patterson and Blackstone, 2022).

REFERENCES

Florea, D., Patterson, W. (2021). A Linguistic Analysis Metric in Detecting Ransomware Cyber-Attacks, www.thesai.org.

Levenshtein, V.I. (1966). February. Binary Codes Capable of Correcting Deletions, Insertions, and Reversals. *Soviet Physics Doklady* (Vol. 10, No. 8, pp. 707–710).

Patterson, W., Blackstone, J. (2022). A Metric to Assist in Detecting International Phishing or Ransomware Attacks. *Proceedings of the 13th International Conference on Applied Human Factors and Ergonomics (AHFE 2022)*, New York, NY.

Simoiu, C., Bonneau, J., Gates, C., Goel, S. (2019). I was told to buy a software or lose my computer. I ignored it: A study of ransomware. In *Fifteenth Symposium on Usable Privacy and Security (SOUPS 2019)*, pp. 155–174.

Section V

Perspectives on Voting Approaches in Cybersecurity

10 Protecting Democracy in an Increasingly Digital World
A Philippine e-Voting Story

William Emmanuel S. Yu

10.1 INTRODUCTION

As the world gets smaller with the use of technology, people expect more and more from it. In the past, we would have to do banking transactions during working hours to allow batch systems to perform end-of-day calculations as systems were not that powerful nor flexible yet. We would also have to perform these transactions in front of a bank teller on the bank premises. We would have to carve a huge chunk of our day to fall in line before these transactions. With advancements in technology, we now perform our transactions anytime, anywhere, and with any device. No more long lines. No more banking hours (24 × 7 banking is now the norm). No more having to carve out a large part of one's day to do banking. On the flip side, we also see more victims of phishing and account takeover attacks. At the end of 2022, digital payments already account for 42.1% of all Philippine financial transactions (Figure 10.1).[1]

This is just one industry and one example. As the world gladly enjoys the benefits of technology in our daily lives, we also see an increase in possible exploits to this. With this amount of money at stake, there will always be bad actors who will attempt to corrupt systems and the technology that power them.

DOI: 10.1201/9781003599142-15

Table 1: Shift and share of digital and non-digital payments BY VOLUME in 2022

		OVERALL	P2X	B2X	G2X
NON-DIGITAL		2,810	1,480	1,328	2
DIGITAL		2,044	1,849	144	51
SHARE OF DIGITAL PAYMENTS	2022	42.1%	55.5%	9.8%	95.9%
	2021	30.3%	36.1%	9.8%	95.3%
	2020	20.1%	23.4%	5.4%	93.2%
	2019	14%	15%	6%	80%

FIGURE 10.1   2022 Share of digital payments in the Philippines (BSP).

No space is more challenging than in democratic elections. In a democracy, the people choose their leaders in free and fair elections. There are an estimated 2.3 billion people in 93 democratic countries, which is roughly 29% of the global population.[2] The Philippines is a country of over 100 million people of which over 65 million are registered voters with over 18,000 positions in the ballots.[3] There are also close to 1.7 million overseas Filipinos who registered to vote. This is a far cry from the over 10 million overseas Filipinos who are eligible to vote. Technology (i.e., Internet Voting) can potentially make it easier for the Filipino diaspora to be part of choosing our leaders. In many countries, technology use has been steadily gaining ground with respect to allowing citizens to vote and in other aspects of the electoral process.

In the latest Asia Foundation study on the state of cybersecurity in the Philippines, it was concluded that there is much to do in terms of improving our information security posture. The report concludes that *the country is still at the initial stage of digital transformation, there seems to be a misconception that threat actors do not pose as serious a threat or that the Philippines is not a target.*[4] This mindset is slowly changing with the increased spate of cybersecurity incidents within and outside our borders. In a previous work,[5] the focus was on misinformation, disinformation, and malinformation in Philippine elections. This time the focus is cybersecurity in the voting technology itself.

This chapter aims to give a view into the challenges of securing e-voting systems. The lens would be the solutions applied in the conduct of Philippine elections. As the official citizens arm for multiple elections, the observations and data used are all collected first hand as election observers. This work ends with a description of some of the gaps and areas to be improved upon.

10.2   THE STATE OF TECHNOLOGY USE IN ELECTIONS

There are many technologies that can be used to support the electoral process. These include support technology such as: improvements in mobile and satellite technology to improve reach and coverage, geographic information systems to improve documentation of voting districts, and biometric technology to improve record capture and voter identification/authentication. The focus on this study would be e-voting technology.

In the latest ICT (Information and Communication Technology) in Election Database of the International Institute for Democracy and Electoral Assistance (International IDEA), out of 193 countries included in the database, there are 29 countries (15.03%) that use e-voting in politically binding national elections and 18 countries (9.33%) that use e-voting in politically binding local elections.

There are many types of e-voting technologies in use by various election management bodies (EMB) worldwide. The most popular ones are:

- **Direct Recording Electronic (DRE)**. This is a form of voting technology that allows voters to directly vote using a digital interface (i.e., touch screen) to select their choices. In many forms of DRE voting systems, the voter is provided with a graphical interface of their choices and the ability to review and correct their votes. Once submitted, generally a voter verifiable paper audit trail (VVPAT) is printed to further allow the voter to review their vote. This technology is used in supervised polling places where voters have to go to cast their vote.
- **Optical Mark Reader (OMR)**. This is a form of voting technology that scans a paper ballot that is marked by the voter. The voter makes their selection on a paper ballot that is subsequently scanned. This generally provides voters with more time to vote when limited machines are available. There is limited ability for a voter to change their votes when this technology is used. The voter generally has the ability to confirm their votes with a VVPAT as well. This technology allows manual recounts. This technology is also used in supervised polling places.
- **Electronic Ballot Printers (EBPs)**. This voting technology provides an interface similar to a DRE where voters can cast their vote. Once the selections are made and confirmed, the device then prints a paper ballot that is placed in a ballot box and subsequently counted. The ballots can be scanned by an OMR device for counting. The OMR devices can be per polling place or centralized. This technology allows manual recounts. This technology is also used in supervised polling places.
- **Internet Voting**. This technology allows voters to cast their votes using an Internet-connected device (i.e., computer or mobile device). Similar to a DRE, the voter is provided with their choices via a digital interface which provides the same ability to check and review their votes before casting them. The use of this technology does not require a supervised polling place and can be done anywhere with Internet access.

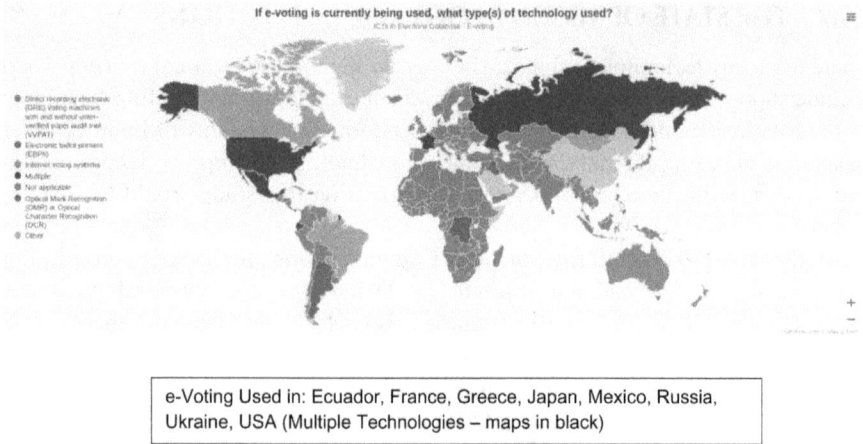

e-Voting Used in: Ecuador, France, Greece, Japan, Mexico, Russia, Ukraine, USA (Multiple Technologies – maps in black)

FIGURE 10.2   Global e-voting technology types in use, 2024. ("e-Voting Technology Use", ICTs in Elections Database, International IDEA, www.idea.int/data-tools/, Last accessed: May 3, 2024.)

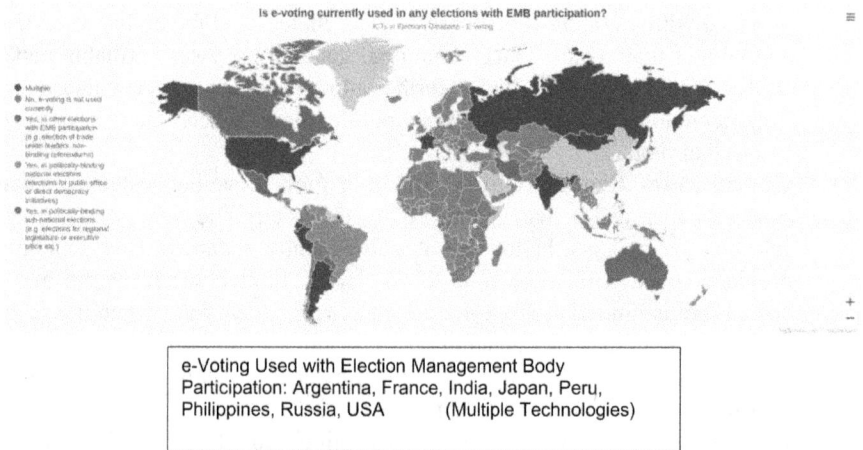

e-Voting Used with Election Management Body Participation: Argentina, France, India, Japan, Peru, Philippines, Russia, USA   (Multiple Technologies)

FIGURE 10.3   Global e-voting adaption 2024. ("e-Voting Technology Use", ICTs in Elections Database, International IDEA, www.idea.int/data-tools/, Last accessed: May 3, 2024.

Each technology provides a three way trade-off between convenience, security, and cost. All of them have their strengths and weaknesses. It is up to the election manager to determine which is suitable for a particular election (Figures 10.2 and 10.3).

10.3   THE STATE OF TECHNOLOGY USE IN ELECTIONS – PHILIPPINES

In the Philippines, we have been using electronic counting and canvassing technology nationwide since the National and Local elections of 2010 (NLE 2010).

NLE 2022 in a Nutshell

	2010	2013	%		2016	%		2019	%		2022	%	
Registered Voters (RV)	51,292,555	52,014,648	1.41%	▲	54,363,844	4.52%	▲	63,559,067	16.91%	▲	65,745,512	3.44%	▲
Clustered Precincts (CP)	76,347	78,166	2.38%	▲	94,279	20.61%	▲	87,540	-7.15%	▼	106,174	21.29%	▲
RV to CP Ratio (Average)	672	666	-0.89%	▼	577	-13.36%	▼	727	26.00%	▲	620	-14.72%	▼
Voting Centers		36,640			29,422	-19.70%	▼	36,807	25.10%	▲	37,144	0.92%	▼
Positions / Contests		18,053			18,083	0.17%	▲	18,072	-0.06%	▲	18,083	0.06%	▲

FIGURE 10.4 NLE 2022 statistics summary.

This was enabled after the passing of RA 9369 or the Automated Election Law on May 14, 2007.[6] The passage of this law laid the groundwork for the use of election technology which includes amending many previous laws involving elections. On August 11, 2008, elections in the Autonomous Region in Muslim Mindanao (ARMM) were used as the pilot for election technology.[7] This regional election was the first time DRE and OMR technologies were used in the country. Since then, the National and Local Elections of 2010, 2013, 2016, 2019, and 2022 have used OMR technology. A few counting machines were also used in a pilot during the last Barangay and Sangguniang Kabataan elections of 2023 (BSKE 2023). OMR technology was preferred as it provides the ability to do a manual audit on the paper ballots. There is a deeper discussion on this under the Random Manual Audit (RMA) later. OMR also requires fewer machines to run the election. In order to further reduce the number of machines needed, the Commission on Elections (COMELEC) grouped existing precincts into clustered precincts. In NLE 2022, there were over 100,000 clustered precincts with each having one OMR device for a maximum of 800 voters and an average of 620 voters per clustered precinct (Figure 10.4).

In addition, RA 10367[8] made biometric capture and registration mandatory for voters in the Philippines. This is an attempt to curb voters who vote multiple times (i.e., flying voters). Under the strength of this new law, the Commission on Elections (COMELEC) implemented the "no bio no boto" policy. This simply means no biometrics, no vote. If one's identity cannot be verified through the biometric records that were captured when the voter registered, that voter cannot vote. This was enforced with the use of biometric verification technology at the polling place during the National and Local Election of 2019 (NLE 2019). The Voter Registration and Verification Machine (VRVM) was used in the elections held on May 13, 2019.[9] The VRVM is a precinct-level machine that is used to biometrically verify a voter before providing that voter with a ballot. Unfortunately, the roll out of the VRVM was not a success and its use was scrapped.

10.4 THE CYBERSECURITY THREAT LANDSCAPE

Increasing use of technology and connectivity have increased the threats we face. A 2018 study commissioned by Microsoft with Frost and Sullivan shows costs of cyber incidents to organizations in the Asia Pacific is potentially $1.75 trillion or around 7% of the regional GDP.[10] In the Philippines alone, it is estimated to be $3.5 billion.

In the last Hiya Global Call Q1 2023 report for the Philippines,[11] we saw 12.1 million nuisance calls and 4.42 million fraud calls. This is potentially 16.53 million chances of being victims. This is just for one country alone. The average Filipino gets four spam calls a month. This is just for spam calls. In this period alone, Hiya observed 6.7 billion spam calls worldwide. This is not limited to fraud related calls and only covers one dimension which is spam calls (not messaging which is larger). There are a lot of more vectors when it comes to use of technology for digital crime. In a three quarter period alone in 2022, Globe (one of the major telecommunications providers in the Philippines) blocked more than 29.3 million[12] (which is around 26% if compared against all mobile subscribers in the Philippines). This is an additional 29.3 million attempts on potential victims.

10.5   OLD THREATS, NEW THREATS – THREATS ON ELECTORAL SYSTEMS

National elections are held every three (3) years in the Philippines. Despite all the safeguards and audit checkpoints that have been put in place, there have been many attempts at gaming the electoral process. Here are some of the more common strategies used.

- **Vote Buying**. This is the general term used when a candidate or political party will provide money or favors in exchange for votes. In a developing country like the Philippines, there are many people who fall for this. Many of our poorer countrymen live under what is called *"Isang Kahig, Isang Tuka"* (one scratch, one peck). The figure of speech pays homage to our fascination with cockfighting (*Sabong*). It simply means a little bit at a time. This reflects the flow of income to many Filipino households, a little bit at a time. So as what limited funds come in they spend it on essentials which is usually food for the family. A windfall brought about by funds obtained from vote buying allows many to live to fight another day (or many days). *Isang Kahig, Isang Tuka.* This is the primary driver and enabler of patronage politics. So when politicians use their position to promote themselves, this can be in the form of selectively providing welfare or jobs to dole outs to potential voters. This is an abuse of state resources to promote a particular politician's self. Locally, this is called *Epal*. Additionally, use of digital technology such as e-money has facilitated its reach and convenience. This takes advantage of the Filipinos strong sense of *Utang na Loob* (debt of gratitude): *Quid pro quo.*
- **Flying Voters**. This is a term used to describe voters who vote multiple times. This can either be in the same location but normally different locations. This is made possible because there are loose and manual controls to prevent voters from voting again. The main control is a placement of a spot of indelible ink in one's finger after voting. Unfortunately, this spot of ink is getting easier

and easier to remove as hygiene technology improves. In other cases, the polling place personnel involved in the process just look the other way. This is also generally used when voters vote on behalf of ghost voters (i.e., dead or nonexisting voters present in tainted or padded voter's list). Probably the term is apt as these voters have to *fly* to make it to as many precincts as possible.

- ***Dagdag Bawas.*** This is when votes are added and/or removed in various stages of the electoral process. Hence the term, *Dagdag Bawas,* which directly translates to Add Subtract. In manual elections (i.e., elections where automation is not used). This can happen during the voting phase, counting phase, canvassing phase, transmission phase (where the certificates of canvass are brought from, let's say, the municipality to the province), and even publication phase. This also has other creative names such as ballot box stuffing. Prior to automation, the precinct level counting and canvassing processes can stretch well into the following day after the election itself. The entire election cycle could take a month. This is why extra care was taken in the past to ensure there are no power outages on election day and subsequent canvassing period to prevent bad actors from carrying out their objectives under the cover of darkness.
- **Disenfranchisement.** This is when a voter is prevented from casting their vote or having their vote counted. This can also be executed in various phases of the electoral process. Techniques such as voter suppression with violence (i.e., firing of machine guns or detonating grenades/mortars just outside polling places, destroying voting paraphernalia) or misinformation (i.e., damaged or relocated polling places) are used. Suppression of voters from the voter's list is also another means of rejection. In some cases, bad actors take ballots and tamper with or destroy them. If that particular precinct is composed of non-supporters (*kontra partido*), then simply detonate a grenade in the immediate area of a polling center during the opening of polls.
- **Controlled Precincts.** This refers to precincts that are controlled by a particular candidate, party, or group. Here no real elections occur and ballots are simply filled up with their desired outcome. Control of these precincts is obtained via the forces of patronage politics, or violence and force. In past cases, ballots were simply filled up by goons and not by the real voters.
- **Digital Election Interference**. This is where threat actors in particular nation states attempt to undermine democracies by interfering with elections. There are many forms of interference with the spread of misinformation, disinformation, and malinformation, especially via social media channels and chat groups, being the most common modes like those in Ukraine and the United States.[13] A concerning scenario is when they attack voting systems and change the outcomes directly.

10.6 IMPACT OF TECHNOLOGY ON THESE THREATS

While technology has the ability to provide a solution to these threats, many of the threats have and continue to evolve. In most cases, threats are minimized and some outright eliminated. There are many controls applied in Philippine elections to combat them. They may be divided into these major areas.

Accuracy. The key, if not the most important aspect to any electoral exercise that makes use of technology, is the accuracy of the results. This is why it is important that there are audit measures in place to validate the integrity of the results. In Philippine elections, there are three (3) major activities that are done to ensure accuracy of election results: Final Testing and Sealing, RMA, and independent third party results verifications (e.g., PPCRV Unofficial Parallel Count).

The *Final Testing and Sealing (FTS)*[14] is a process in which voting machines are tested a few days before the election in full view of election stakeholders and observers to ensure that proper counting is done by the machines. During the FTS, a simulated election session is opened and a few ballots are cast by the participants. Afterwards, the election is closed and the election results (ER) are generated (vote tally). The ballot boxes are also opened and each ballot is manually counted in full view of all of the stakeholders who participated. The results should match. Once completed, the machines are re-zeroed and sealed. These are opened again only on election day. It is important that this entire proceeding is done in full view of observers and that they maintain the proper chain of custody. This is done for all the clustered precincts nationwide.

The *RMA*[15] is an audit process that is executed after the election. It is similar to the FTS but applied to all the ballots in the ballot box after actual elections close. However, doing this in over 100,000 precincts would take a long time. Instead, a sampling from one to three clustered precincts per congressional district is randomly selected and subjected to the RMA. The number of clustered precincts subject to audit is proportional to the population of that congressional district.

At the start of the RMA, each ballot box from the selected precincts is opened in front of observers and the votes recorded on each ballot are manually counted. Results are then compared against the results generated by the counting machine. For the 2022 NLE, the RMA committee (RMAC) was composed of representatives from the COMELEC, Philippine Statistics Authority (PSA), and a Coalition of Civil Society Organizations (CCSOs) that included the Legal Network for Truthful Elections (LENTE), the National Movement for Free Elections (NAMFREL), the Philippine Institute of Certified Public Accountants (PICPA), the Philippine Society of Public Health Physicians (PSPHP), and the Information Systems Audit and Control Association (ISACA). Volunteers from PPCRV were also present to observe and support the activities of the RMA. For NLE 2022, the RMA overall accuracy for audited positions is 99.958%.[16]

The *Parish Pastoral Council for Responsible Voting (PPCRV) Unofficial Parallel Count (UPC)* is a process that aims to ensure there is no electronic *Dagdag Bawas*.[17] The PPCRV is an official Citizen's Arm of the Commission on Elections. It has a three-fold mandate: voter's education, poll watching, and performing the Unofficial Parallel Count (UPC). The UPC is essentially an audit activity where poll watchers in every clustered precinct nationwide observe the proceedings of the election to ensure that everything is in order. At the end of the elections, a total of up to 32 election returns (ERs) are printed; eight (8) copies prior to transmission of the results from the VCM to the Municipal Board of Canvassers (MBOC), COMELEC server and Transparency Server, and twenty four (24) after transmission. A PPCRV poll watcher obtains one of the pre-transmission copies of this ER and these are sent to a centralized command center. In NLE 2022, the UPC was carried out in the Quadricentennial Gym of the University of Santo Tomas (UST) in Manila. These ERs nationwide are subject to a double blind encoding process and compared against the transmitted electronic results that were sent from each of the VCMs to the Transparency Server. The results should match as well. This is done for as many clustered precincts with both electronic and manual returns. For NLE 2022, 89,260 ER have matching manual and electronic returns. This represents 82.81% of all returns and 99.848% of all returns with both electronic and manual ERs matched. About 134 mismatched returns were flagged for validation due to either incomplete or damaged/unreadable manual returns. The 10 mismatches have been escalated for further validation and investigation (Figure 10.5).

Speed. The speed in which counting and transmission can be achieved with the use of technology has narrowed the potential window for fraud, particularly *Dagdag Bawas*. This applies to both counting and transmission. In the case of counting, the most common type of fraud is modification of the vote tally after elections at the polling place. This can be done via intimidation, bribes, and other techniques. There is a saying here that one may lead the count but when there is a power failure he has now lost. *Llamado siya pero isang brown-out lang talo na*. This is mitigated as counting machines are tested and sealed (i.e., FTS), and counting is done in front of poll watchers. The counting, transmission, and canvassing process is fully automated. This is why we call our system an automated election system (AES). This was done to prevent tampering of votes and manual intervention during counting. Transmissions are protected using industry best practices in information security such

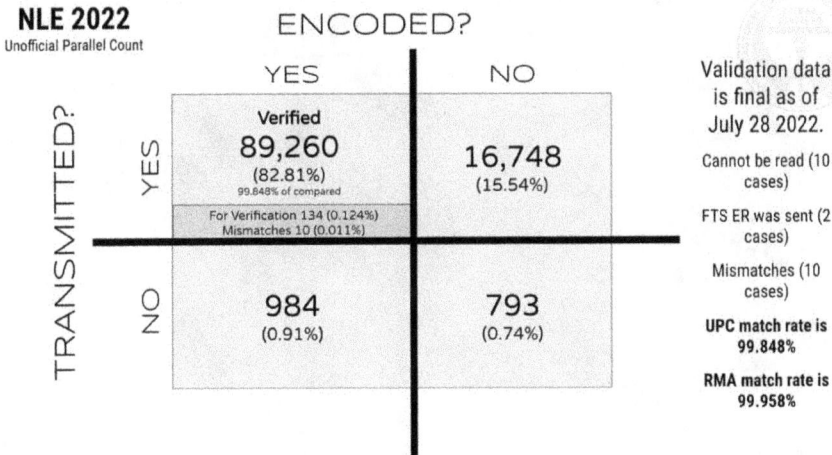

FIGURE 10.5   Summary of PPCRV UPC results for NLE 2022.

as the use of industry grade cryptography (providing confidentiality, integrity, and nonrepudiation), and the wide distribution of tally results. Transmission rates have been steadily improving due to advancements in data transmission technology and improvements in network coverage. In NLE 2022, 35.22% of results came in one (1) hour of closing and 90.83% of results within seven (7) hours of closing. Before the night is over, the results are already known. This narrows the window for potential *Dagdag Bawas* (Figure 10.6).

Integrity. Ensuring integrity of results is important in elections. There are two general mechanisms applied to ensure this.

(a)

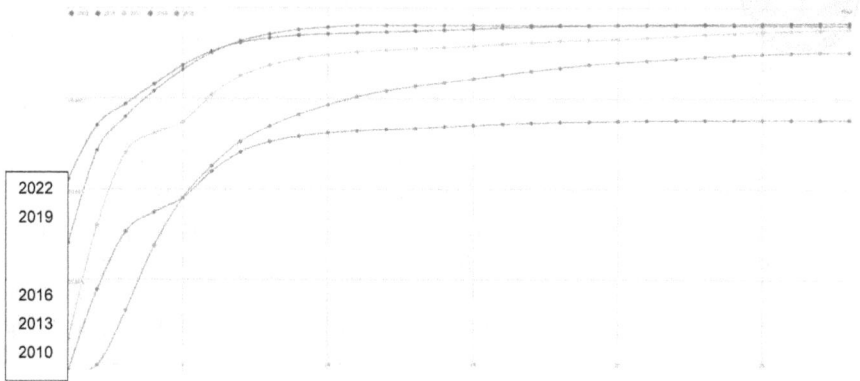

Transmission Rate Comparison for 2022 and Previous NLEs

(b) Transmission Rate Comparison for 2022 and Previous NLEs

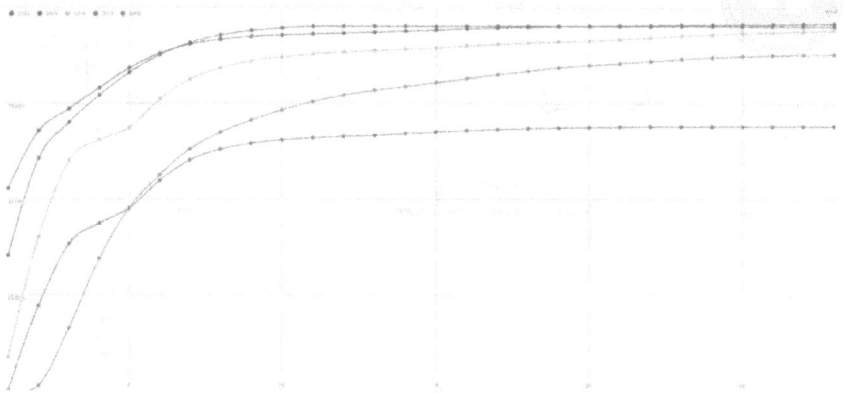

FIGURE 10.6 Transmission rate comparison between NLE 2010, 2013, 2016, 2019, and 2022.

- **Use of Cryptography**. There are multiple layers of cryptography used in the election. There are both multiple layers of payload and transport encryption being applied. There is also an additional step of digitally signing election results by the Electoral Board (EB) members using Public Key Infrastructure (PKI) technology. All cryptography used in the elections thus far are based on industry standards.
- **Wide Distribution of Results**. Since election results are public information anyway, there is no requirement to ensure privacy of results. Therefore, distributing the results of the election as widely as possible also provides a form of protection of its own since any form of subsequent tampering will be very apparent as there are multiple copies to compare against. This is why there is public access to election results via the COMELEC public access website and election returns posted prominently on polling places.

Another aspect of integrity is the voting systems themselves. There are multiple mechanisms applied to ensure integrity.

- **Local Source Code Review (LSCR)**. This is where the source code of the election system is subject to public review by qualified representatives of the different political parties, civil society organizations, and the academe. Observers can apply to be source code reviewers. The rules with respect to the LSCR have changed throughout the different election cycles but have generally become more open. There was no LSCR for NLE 2010. There was a short one after the elections for NLE 2013. There was a long four (4) month one for NLE 2022.
  After the source code is reviewed, the entire system is built in a public **Final Trusted Build** ceremony. All components are properly documented and hashed to allow subsequent verification. After this, a copy is used as the build master for loading into the voting machines. Another copy is deposited in the Banko Sentral ng Pilipinas (BSP) vault, the Philippine Central Bank.
- **International Certification Entity (ICE)**. This is a third-party independent audit of the election system that covers multiple aspects including physical tests, accuracy tests, software tests, security tests and functionality tests. A number of ISO standards, the US Election Assistance Commission (EAC) Voluntary Voting System Guidelines (VVSG) adapted to the requirements of RA 9369 are used as the basis of the audit. The last election was audited by ProV&V of Huntsville, Alabama, USA.
- **Hardware Acceptance Testing (HAT).** This test involves testing the voting machines. This is performed to focus on the scanning hardware. This test is generally performed by the COMELEC.
- **Pre-Logic and Acceptance Testing (Pre-LAT).** This is a full end-to-end software and hardware test. Here the production software that was built is loaded into the voting machines and subject to an end to end from opening, voting, closing, and transmission. There are mock ballots and mock canvassing servers used during the test. This test takes multiple months to complete and is executed on all the voting machines before shipping to their designated

polling centers. This test is performed by the COMELEC and has been made open to observers in the past few elections.

- **Technical Evaluation Committee (TEC) certification.** This group is composed of representatives from the COMELEC, Department of Science and Technology (DOST), and Department of Information and Communications Technology (DICT). They take the results of the activities above together with additional activities and tests and provide a certification for the system to be used in an election.

- **Final Testing and Sealing (FTS).** Another purpose of the FTS described previously is to ensure that machines work when they get to the polling place. No election can open without the machines being subjected to an FTS.

Auditability and Transparency. It is estimated that the amount of paper trail produced to document Philippine elections could circle the globe twice. This does not consider the number of digital copies of the election data that are stored in various systems including those of the election observers and political parties. The sheer amount of data shared and distributed in the national and local elections in the Philippines is quite a lot. The goal is to ensure that many people get as many copies of the data as it is made available. This provides a substantial paper and digital trail to allow for independent verification and audit.

- **Paper Ballot**. There were over 65 million voters in the NLE 2022 and there are potentially 65 million paper ballots that could be marked when voted upon. This provides the ability to verify and audit results. One of the mechanisms where Paper Ballot is obviously crucial is during the RMA. The paper ballots are also kept and secured for potential electoral protests in the city or municipal treasurers office in the event of a recount.

- **VVPAT**. This is a small piece of paper that is printed after a voter casts their vote. This allows the voter to immediately review their vote to see if it matches what they cast. Under the current procedures, a voter can immediately note a protest in the minutes of the clustered precinct if they find a discrepancy between what was printed and what they remember writing on the ballot. This can flag a clustered precinct for potential investigation later on. While there were a few cases reported in the media, so far, no such formal complaint has been filed. The VVPAT is dropped into a box which is kept together with the ballots. This can also serve as another tool for post-election audits. There should be as many VVPATs as ballots. There is currently a proposal to include VVPATs in the RMA audits in future elections.

- **Printed Election Returns (ER)**. This refers to the actual election returns that are printed after an election. This is also referred to as physical ERs. There are two classes of ERs. There are the pre-transmission ERs and post-transmission ERs. As the name implies, the pre-transmission copy of the ERs are printed before network transmission. This ensures that there is no transmission before this point and the voting machines are not connected to any external communications network until after the first eight (8) pre-transmission copies of the ERs are printed. The pre-transmission copies are generally given to the municipal or city board of canvassers (MBOC/CBOC), Congress, COMELEC, accredited citizen's arm (PPCRV), dominant majority party, dominant minority party, ballot, and one for posting in the polling place. After transmission, other accredited citizen's arms, media, political parties and observers can obtain copies of the post-transmission ERs. Together with the digital ERs, this is used by organizations such as the PPCRV for independent verification of results (i.e., UPC).

- **Transparency Server and the Digital Election Returns (ER).** As part of the election process, results are transmitted in a ladderized fashion for canvassing. In general, municipal positions are canvassed first then provincial then national. Each OMR device transmits the results three (3) times to the Transparency server, COMELEC server, and MBOC/CBOC canvassing server. The typical canvassing path for the election results is from the clustered precinct vote counting machine (VCM) to the Municipal or City Board of Canvassers (M/CBOC) then to the Provincial Board of Canvassers (PBOC) then to the National Board of Canvassers (NBOC) (Figure 10.7).

Results at each canvassing stage is what is used to proclaim winners at the various stages of the election. The COMELEC server serves as the backup server and also feeds the data it receives to the COMELEC public access website. The M/CBOC canvassing server is used to canvas their respective results before forwarding the canvas to the next level which is the provincial board of canvassers (PBOC) canvassing server. This is used for proclamation. The Transparency server is where accredited citizen's arms, political parties

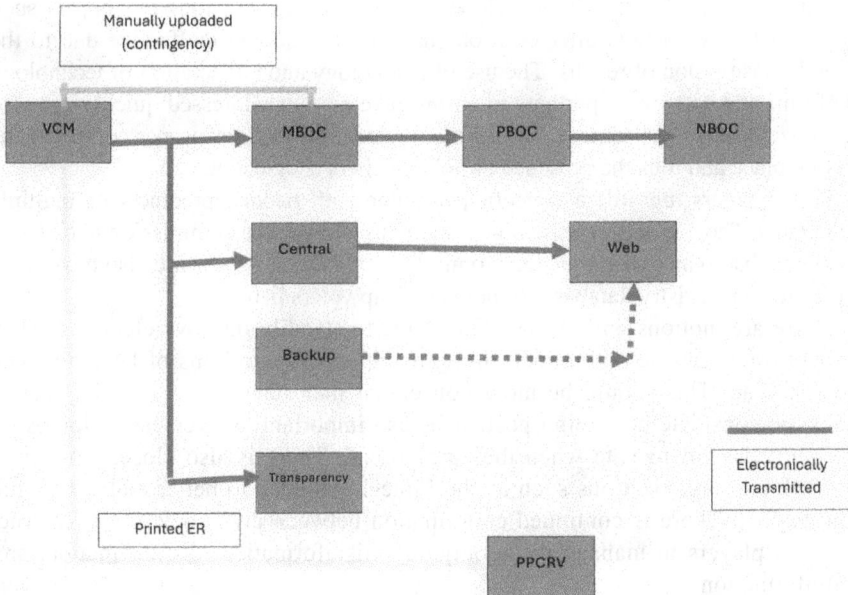

FIGURE 10.7 NLE election return transmission flow.

and media get a digital copy of the ERs. Together with the physical ERs, this is used by organizations such as the PPCRV for independent verification of results such as the Unofficial Parallel Count (UPC).

- **COMELEC Public Access Website**. This is accessible over the Internet to the public. Here anybody can get a copy of the election results and can drill down to get precinct-level election results.

It is hoped that the sheer amount of physical and digital copies of the results will make it difficult to tamper with the system. It is designed to ensure that the most number of people get the results at the soonest possible time.

10.7   SOME THREATS LEFT UNRESOLVED AND CHALLENGES AHEAD

Despite the use of technology and many controls applied, there are still open threats. Many of these are outside the realm of voting systems.

The main challenge is still the scourge of vote buying. There have been many attempts to curb this such as large cash withdrawal limits, limits on use of government funds during the campaign period, election spending monitoring, e-money monitoring, and many others. A few candidates have already been disqualified because of violations. It is hoped that this will serve as a strong disincentive on these vote buying practices. Anecdotally, it has been observed that there is a lot more vote buying ongoing with higher amounts in the last few elections.

There is also the threat of disenfranchisement and controlled precincts. The use of violence and force cannot be eliminated fully and requires the assistance of law enforcement or the military. However, some of the previous techniques such as destruction of ballots after elections have been rendered ineffective due to the digital transmission of results. The use of technology and cooperation of technology platforms and ecosystem partners to ensure fake news is addressed quickly can also minimize disenfranchisement. The current system still requires proper control of the polling place and thus the presence of poll watchers is essential.

Flying voters are still a concern particular with padded precincts or multiple registrants. The use of biometric registration has allowed the commission to clean up the voters list to an extent that aims to minimize this issue. There are also motions to integrate civil registry databases to help clean up voters lists.

There are motions within the COMELEC to steadily improve elections. They are exploring the use of Internet Voting to support the millions of Filipinos who work abroad. This should be more convenient than having to go to the nearest embassy, consulate or foreign post. It is also important as overseas Filipinos do not get a day off to vote when they are abroad. There is also closer work with the BSP and organizations such as the Fintech Alliance to better address digital vote buying. There is continued coordination between civil society and Internet platform players to manage the scourge of misinformation, disinformation, and malinformation.

The threat of the 3Gs: Guns, Goons, and Gold is still ever present today. The proper use of technology has hopefully reduced the worst of the violence that historically accompanies elections in the Philippines. As long as election violence exists, there is work to be done. Credible elections are the key to a credible democracy. In the meantime, organizations such as the PPCRV, political parties, and other election stakeholders must stay ever vigilant striving to ensure that we have Clean, Honest, Accurate, Meaningful and Peaceful (CHAMP) elections.

NOTES

1   Banko Sentral ng Pilipinas. 2022, "2022 Status of Digital Payments", Banko Sentral ng Pilippines Report.
2   Bastian Herre, Lucas Rodés-Guirao, Esteban Ortiz-Ospina and Max Roser. 2024, "Democracy", Our World in Data, https://ourworldindata.org/democracy.
3   Commission on Elections, 2022. "Project of Precincts", Philippine Commission on Elections.
4   Secure Connections. 2022, "Cybersecurity in the Philippines: Global Context and Local Challenges", The Asia Foundation – Philippines.
5   William Yu. 2024, "Studying Fake News Proliferation by Detecting Coordinated Inauthentic Behavior", *New Perspectives in Behavioral Cybersecurity*, Routledge.
6   "RA 9369: AN ACT AUTHORIZING THE COMMISSION ON ELECTIONS TO USE AN AUTOMATED ELECTION SYSTEM IN THE MAY 11, 1998 NATIONAL OR LOCAL ELECTIONS AND IN SUBSEQUENT NATIONAL AND LOCAL ELECTORAL EXERCISES, TO ENCOURAGE TRANSPARENCY, CREDIBILITY, FAIRNESS AND ACCURACY OF ELECTIONS, AMENDING FOR THE PURPOSE BATAS PAMPANSA BLG. 881, AS AMEMDED, REPUBLIC ACT NO. 7166 AND OTHER RELATED ELECTIONS LAWS, PROVIDING FUNDS THEREFOR AND FOR OTHER PURPOSES", Senate and the House of Representatives of the Philippines in Congress, May 14, 2007.
7   Edwin Fernandez and Kristine Alave. 2008, "High-Tech Polls: No More 'Hello Garci' ", *Philippine Daily Inquirer*, August 11, 2008.
8   "RA 10367: AN ACT PROVIDING FOR MANDATORY BIOMETRICS VOTER REGISTRATION", Senate and the House of Representatives of the Philippines in Congress, July 23, 2012.
9   Shiela De La Cruz. 2019, "72 Percent of VRVM in Cotabato City, Maguindanao Fail to Work", *Philippine Daily Inquirer*, https://newsinfo.inquirer.net/1118476/72-percent-of-vrvm-in-cotabato-city-maguindanao-fail-to-work May 14, 2019.
10   Frost and Sullivan. 2018, "Understanding the Cybersecurity Threat Landscape in Asia Pacific: Securing the Modern Enterprise in a Digital World", Frost and Sullivan, https://news.microsoft.com/apac/2018/05/18/cybersecurity-threats-to-cost-organizations-in-asia-pacific-us1-75-trillion-in-economic-losses/, 2018.
11   Hiya Global Call Threat Report Q1 2023. Hiya.
12   Globe Telecoms. 2023, "Globe Sees 85% Decline in Scam, Spam SMS, Credits Strong Partnership with Banks Against Fraud", Globe Telecoms Corporate Release, www.globe.ph/about-us/newsr oom/corporate/globe-partnership-with-banks-vs-fraud-sees-85-percent-decline-in-scam-spam-sms-blocked-as-of-q3-of-2023#gref.
13   Adrian Shahbaz, Allie Funk. 2019, "Freedom on the Net 2019 Key Finding: Politicians and Hyperpartisans Use Digital Means to Manipulate Elections", Freedom House, https://freedomho use.org/report/freedom-on-the-net/2019/the-crisis-of-social-media/digital-election-interference, Last accessed: May 3, 2024.
14   "Resolution No. 10727 – General Instructions For The Constitution, Composition And Appointment Of Electoral Boards; The Process Of Final Testing And Sealing Of The Vote Counting Machines; And The Voting, Counting And Transmission Of Election Results In Connection With The 09 May

2022 National And Local Elections", Philippine Commission on Elections Resolutions, November 10, 2021.

15   "Resolution No. 10738 – In the Matter of the General Instructions for the Conduct of Random Manual Audit (RMA) for the 09 May 2022 Automated Synchronized National and Local Elections and Subsequent Elections Thereafter", Philippine Commission on Elections Resolutions, December 9, 2021.

16   "NAMFREL Final Report, 2022 National and Local Elections", National Citizens' Movement for Free Elections, 2022.

17   William Yu and Rommel Bernardo. 2022, "PPCRV Final IT Report for NLE 2022", Parish Pastoral Council for Responsible Voting.

11 Passwords

An Empirical Study of Behaviors in the Social Media Era

Augustine Orgah, Sneha Sudhakaran, and Dyaisha Orgah

11.1 INTRODUCTION

Credentials are a common means of authenticating or gaining access to a system much as a physical key is to a lock. Credentials could be a physical token, biometric information, a password, a passphrase, unique identifiers that should not be duplicated but specific to a user, or a combination of the aforementioned. For this project, credentials will mean an account username/identifier and password combination with a focus specifically on passwords.

A password is personal, it could be something a person knows, can identify, can recall, a set of unique characteristics of a person or involving various things. Passwords are the most common means of authentication in the digital space [1]. A password should be secure, unique, and follow best practices such as exhibiting complexity and having sufficient length [2, 3]. The longer a password is, the stronger it is and less susceptible to guessing, dictionary, or brute-force attacks. A minimum length of 16 characters is desired as it yields less predictable passwords [4–6].

Typically, passwords are in text form containing characters from a class or group of classes such as roman numerals and symbols, the English alphabet, and so on. Passwords are the most common mechanism for authenticating humans to computer systems [5].

DOI: 10.1201/9781003599142-16

Have password behaviors changed or remained the same in the last 10 years or more? Are there any factors affecting password behaviors in 2024? Are there knowledge/technical gaps? Due to an increasing digital society, the average user is more technologically literate or predisposed to technology. Does this mean that stronger passwords are in use today? Are security behaviors enhanced in 2024?

It is safe to say that in today's increasingly digital world, the average user has multiple credentials (username and password) that range across domains that are personal, professional, educational, medical, entertainment and leisure, financial and more. Credentials are now needed to order food, buy groceries, drive cars and access subscription services or features, and lots of IoT (Internet of Things) devices and more.

The average user may need to manage upwards of 15 credentials and keep them safe. For instance, a college student in the USA will have credentials for school and personal email, learning management systems, gaming console account, devices (laptop, tablet, smart phone), cloud storage, social media accounts, streaming platforms, financial accounts, and food ordering services to name a few. How are these credentials managed? Are there healthy security behaviors for these credentials pre and post creation?

Our intention with this study is to gain insights and report on the observations about current affairs regarding password behaviors in the highly digital era of social media via subjects surveyed.

11.2   THE STATE OF AFFAIRS

According to IBM's 2023 data breach report, 15% of breaches were due to compromised or stolen credentials. Phishing attempts and business email compromise are the popular attack vectors leading to obtaining credentials [7]. The techniques identified by IBM data breach report and other methods rely on stolen or compromised passwords to be effective. Methods commonly used, but not limited to by malicious actors include [3, 4, 8, 9]:

- Brute Force – one that aims to use all attempts/combinations within the search space of the password until the right match (the password) is found.
- Password Spraying – an attack that utilizes a password across multiple accounts. This is a type of brute force attack that analogously uses a single key(password) to try to open multiple doors/accounts.
- Phishing – an attack that tricks or manipulates a victim into providing their password and other personal information while the user thinks they are performing legitimate actions. Usually occurs via email and utilizes some form of social engineering.
- Dictionary Attack – one in which unacceptable, common, and trivial words are used as passwords and are usually contained as part of curated password lists, for example, 12345, abc123, etc.
- Keylogging – an attack in which the keystrokes of a victim are recorded, stored, and often transmitted to a malicious actor without the knowledge and permission of the victim.

- Credential Stuffing – an attack that utilizes stolen credential pairs of a username/account ID and password combination across login forms in an automated fashion to gain access.

In October 2023, genetics and ancestry discovery company 23andMe fell victim to a credential stuffing attack because passwords of clients that were used on 23andMe.com were also used on other compromised sites [10]. In May 2021, due to a compromised password that was part of a previous breach, bad actors gained access to the US fuel Colonial Pipeline network, ransomed, and exfiltrated data. The Colonial Pipeline network breach occurred because of an exposed password to a virtual private network (VPN) account that was reused elsewhere [11, 12].

Most users have password behaviors that may seem innocuous and safe from bad actors. However, according to Gaw and Edwards, computer users perceive computer attacks or compromises against them to emanate from people close to them, who are the most motivated and capable [13]. It is commonly accepted that as security increases, convenience, or even user-experience/usability decreases. Attaining a balance between security and convenience can be a slippery slope. Due to the security–convenience tradeoff, users tend to engage in password-management behaviors that are not desirable because they do not foresee any negative consequences to themselves or others immediately [14]. The seemingly innocuous behaviors that could lead to breaches, theft, unavailability of services/systems, ransoms, and even death are not limited to sharing passwords, reusing the same passwords in multiple places, oversharing of information on social media, indiscriminate clicking of links, visiting malicious/suspicious websites, accessing and opening documents from an untrusted source via email or from external media, poor software update cadence, and poor physical security of electronic devices [15].

11.2.1 Password Management and Modification

We consider password creation, storage, and transfer as core password management processes. Poor password management practices could have dangerous consequences for an individual, organization, larger communities, and beyond.

Consider the password creation requirements for AT&T's Smart Home Manager [16] account. Passwords must use:

- 8 to 24 characters
- An uppercase letter
- A lowercase letter
- A number
- One of these special characters + = #? *$! – _

A compliant password for an AT&T account will have complexity, contain four distinctive character classes, and have a minimum length of 8 and maximum of 24 characters. Research has shown that the increase in character class requirements for a password decreases the ability to guess the passwords, but it also causes frustration and difficulty for users [6]. This frustration and difficulty due to complexity can lead to other behaviors such as reuse, sharing, and difficulty with memorization leading to potentially vulnerable situations whereby passwords are written down or stored in an unsafe manner electronically [3, 17].

Komanduri et al. have identified that most users will write down or store their passwords. They found that the likelihood of this behavior was related to complex passwords [5]. Due to the complexity of passwords and their requirements, once users generate a good password – one of good length, complexity, and void of dictionary words, they tend to keep it or modify it for reuse. Woods and Mikko find that users make minor changes to their passwords, such as adding or removing a number or a special character. They also highlight that a user's future password can be guessed easily due to a user's modification behavior [17].

An easily guessed password is neither good nor is it complex. It means the password is highly likely to be a dictionary word or part of a curated password list or one that has been released as part of a breach. According to the National Institute of Standards and Technology (NIST), humans, "have only a limited ability to memorize complex, arbitrary secrets, so they often choose passwords that can be easily guessed" [3]. A 2024 global password survey by Bitwarden revealed that 36% of the respondents confirmed using personal information in their passwords that can easily be found on social media [4].

As it pertains to complexity and length of passwords, keystroke logging, phishing, and social engineering attacks are equally effective on simple passwords as they are on complex ones because they are not affected by password length and complexity [3]. AT&T on March 30, 2024, stated that data specific to AT&T was part of a data set released to the dark web containing personal identifiable information of current and former customers. Investigations are still ongoing to identify how data was exfiltrated via AT&T or its vendors [18].

11.2.2 PASSWORD SHARING

Apart from writing passwords down due to complexity or difficulty with memorization, users also share their passwords with others. According to Merdenyan and Petrie and a survey on password sharing strategies conducted with 122 respondents from a range of countries, they highlight that more than 30% of the respondents shared their email password, and 25% shared their Facebook password with close friends or partners [1]. Netflix Inc. on February 8, 2023, reported that over 100 million households were sharing accounts impacting their business [19]. The sharing of passwords with friends and family, or with strangers, can have devastating consequences and is a prevalent example of poor password hygiene and management [15].

11.2.3 PASSWORD REUSE

Password reuse, using the same passwords in multiple places, is a common factor leading to computer breaches today [5, 15, 20]. A 2024 global password survey by Bitwarden revealed that 25% of respondents admitted to reusing passwords across as many as 20 accounts [4]. Password reuse is prevalent and often utilized as a strategy and means of remembering multiple passwords [1, 17] and for easier management. The danger of this practice is that the exposure of one password can lead to the breach of several accounts

[13]. Merdenyan and Petrie reported that 94% of the participants in the questionnaire reported reusing at least one password for more than one system. Password reuse was increasingly to occur with the more online accounts participants owned [1].

11.2.4 PASSWORD LENGTH

Research shows that the longer a password is, the better. Password length is the primary factor in determining password strength. For instance, a 16-character password will yield significantly less predictable passwords than a 6-character one. Passwords that are too short yield to brute force attacks as well as to dictionary attacks using words and commonly chosen passwords [3–5].

Previous studies have shown that about 30% of users chose passwords whose length is equal to or below six characters. Data from the top 10 and top 50 common password lists from 2024 show that 50% and 42% of passwords on the list have six characters or less [4, 8, 21]. Research shows that increasing minimum length requirements may increase password strength more than relying just on character class requirements [6].

11.2.5 BAD PASSWORDS

Passwords that are part of list of commonly used, expected, breached corpuses, or contain personal identifying information are bad and should be avoided [2, 3, 6]. In 2009, a major password breach led to the release of 32 million passwords from which the password *123456* emerged as the most common [8]. In 2024, almost 15 years later, two separate reports list the password *123456* as the first password on their top 10 and top 50 password lists of 2024 [4, 8, 21]. Analysis from the 2009 data breaches revealed that 60% of users had passwords that qualified as bad passwords. Almost half of the users used slang, dictionary, trivial or common words, consecutive digits, adjacent keyboard keys, etc. [8].

Why are passwords such as *123456* or *123123* considered bad? Besides the fact that they are part of most password lists, dictionary words, and can be easily guessed, they are susceptible to brute force attacks. For instance, consider the target password *123456*, the search space for all combinations consists of only numbers, 0–9. This means a search space of 10 characters to choose from per digit for each character in the 6-digit password. Therefore, $10 \times 10 \times 10 \times 10 \times 10 \times 10 = 10^6$ or 1,000,000 possibilities/combinations. The number of combinations can thus be represented by the following formula

$$\text{Combinations Possible} = \text{Number of Characters}^{\text{Length of Password}}$$

According to Gibson Research Corporation, the haystack (search space) for this needle (*123456*) will require approximately 19 mins to exhaust all combinations if the computer guessed 1000 guesses per second. It will take 0.0000111 seconds assuming 100 billion guesses per second [22]. In 2012, a computer was built to perform up to 350 billion guesses per second [9, 23]. Hashcat, a popular, opensource, and fast password cracker can also perform billions of guesses per second [24]. Consider the

8-digit password *Aa123456* which is also part of the top 50 most common passwords in 2024, the haystack (search space) consists of 26 uppercase letters (A–Z), 26 lowercase letters (a–z), and 10 digits (0–9). This means a search space of 26 + 26 + 10 = 62 characters and $62^8 = 2.2 \times 10^{14}$ combinations. Given a computer capable of one thousand guesses per second, it will take 70.56 centuries to exhaustively search all the possible combinations and approximately 37 mins if the computer performs a hundred billion guesses per second [22]. Both passwords are shown to be prone to brute force attacks given computational resources. They are also common and part of dictionary lists or curated password lists which means they do not necessarily have to be guessed and could be looked up even quicker.

We have presented some of the reported and researched password behaviors of users in years past and aim to accomplish the same with our study albeit in what we consider as a different era, the social media era. One of the questions we hope to unravel with our study in 2024 is to find out if the trends and tendencies from the last 10 years remain true today and what are the modern trends observed in today's ever-increasing digital space?

11.3   METHODOLOGY

The contents of Table 11.1 constitute the survey questions provided to all participants via a URL. Before completing the survey, all participants must select an option between voluntarily completing it or exiting immediately if they did not. A part of the disclaimer, participants are notified of the complete anonymity, voluntary nature, and absence of collection or request of any personal identifiable information. 151 participants reviewed the disclaimer and consent message, with 7 declining to partici-pate in the survey, leaving 144 participants who opted in and completed the survey. This survey is also cleared for human participants and designated with an IRB exempt review status and IRB Number 24-170.

Two short answer questions are excluded for brevity. They are:

- What is your reasoning behind your answer to Question 2?
  - Do you use a password manager for storage/safeguarding of your personal password(s)?
- What is your reasoning behind your answer in Question7?
  - How many of your personal accounts have the same password?

Age of respondents (Years)

<18: 23%; **18-25:** 21%; **26-35:** 11%; **36-45:** 20%;
46-55: 14%; **>55:** 9%; **Prefer not to say:** 2%

Industry of employment of respondents

Education: 17%; **Government:** 6%; **Financial:** 4%; **Public service:** 10%; **IT (Information Technology):** 10%; **Manufacturing:** 2%; **Student(unemployed):** 36%; **Other:**11%; **Undisclosed:** 4%

TABLE 11.1
Survey Responses

#	Question	Responses				
1	Will you assert that your behaviors regarding your personal passwords are healthy? (This means good complexity, length, storage, nonreuse, etc.)	Extremely healthy: 18%	Somewhat healthy: 47%	Neutral: 10%	Not very healthy: 19%	Not healthy at all: 6%
2	Do you use a password manager for storage/safeguarding of your personal password(s)?	Yes: 33%	No: 47%	Sometimes: 20%		
3	Do any of the following prohibit you from using a password manager?	Every time: 7%	Very often: 11%	Sometimes: 13%	Seldom: 17%	Not at all: 53%
4	Do any of the following prohibit you from using a password manager?	Cost: 8%	I do not need/want one: 17%	I need to learn how to use one: 13%	I do not trust them: 13%	Too many choices to pick from: 8%
5	Do you come up with a new password every time you are required to create or reset your personal password(s)?	Every time: 19%	Very often: 20%	Sometimes: 28%	Rarely: 24%	Never: 9%
6	Are you likely to reuse a password if allowed to do so?	Very likely: 44%	Likely: 24%	Occasionally: 16%	Seldomly likely: 11%	Never: 5%
7	How many of your personal accounts have the same password?	0: 14% 1-3: 19%	3-5: 25% 5-9: 19%	10 or more: 19% Other: 3%		
8	Does the regularity of password change requirements influence password reuse?	Always: 9%	Very often: 28%	Sometimes: 41%	Seldom: 16%	Never: 6%

(continued)

TABLE 11.1 (Continued)
Survey Responses

#	Question	Responses
9	How often do you change passwords to your personal accounts?	Monthly: 4% Quarterly: 9% — Never: 38% Every six months: 4% — Yearly: 14% — Other – Mixed: 31%
10	Do you change the default passwords on electronic devices or accounts you own?	Always: 44% Not important to me: 6% — Sometimes: 31% Never: 13% — I usually need assistance with doing so: 3% — I cannot change default password because it is permanent: 1%
11	Do you share passwords to accounts you own or manage?	Always: 1% — Very often: 8% — Sometimes: 20% — Seldom: 30% — Never: 42%
12	Have you shared your personal passwords to any of the following categories of accounts?	Digital streaming platforms (Netflix, Hulu, Spotify, etc.): 20% — Email: 5% Banking/Financial: 3% — E-Commerce (Amazon, Walmart, Etsy, etc.): 3% — I do not share my password(s) with anyone: 42% — Other-mixed: 27%
13	Do your passwords/passphrases contain personal information related to you? E.g., Name of pet(s), birthdays, zip codes, etc.	Yes: 26% — No: 47% — Sometimes: 28%
14	Do you use non-English words/phrases in your passwords?	Yes: 22% — No: 61% — Sometimes: 17%
15	Do you store passwords in one or more of the following ways?	Plain text file on an electronic device: 11% — Handwritten in a book/paper or other stationary medium: 18% — In the cloud: Google, Office 365, iCloud, etc.: 7% — In browsers for use via autofill: 14% — Other-mixed: 50%

No.	Question					
16	How challenging is it for you to generate a good, strong password? One with good length and complexity	Very challenging: 15%	Challenging: 23%	Neutral: 25%	Somewhat challenging: 10%	Not at all challenging: 28%
17	Do you consider ji32k7au4a83 to be a good password?	Yes: 31%	No: 48%	Maybe: 22%		
18	Do you consider the following passwords the same? password123 pAssWord123 123p@$$w0rD 1pass3word5	Yes: 22%	No: 78%			
19	Do you consider the following passwords the same? Fall2024! Fall2025!	Yes: 60%	No: 40%			
20	On average, what is the length of your passwords?	They contain 16 or more characters: 24%	They contain less than 16 characters: 76%			
21	Do you have passwords that meet the following?	contain 8 or less characters: 26%	contain at most 10 characters: 58%	Mixed: 16%		
22	Do you have two-step verification or multi-step verification enabled across personal accounts you own?	Yes: 33%	Very often: 19%	Sometimes: 34%	Rarely: 4%	No: 10%
23	Do you know what 2FA means in relation to accounts?	Yes: 80%	No: 14%	Somewhat: 6%		
24	Do you know what MFA means in relation to accounts?	Yes: 56%	No: 32%	Somewhat: 12%		
25	Some services such as online banking require users to sign up for MFA. In cases where it is not required, what is the likelihood you would enable it?	Very likely: 35%	Likely: 24%	Neutral: 21%	Unlikely: 18% Never: 2%	I do not know what MFA is or does: 11%

(continued)

TABLE 11.1 (Continued)
Survey Responses

#	Question	Responses			
26	Do you use sites such as [https://haveibe enpwned.com/Passwords] to verify your passwords or credentials being part of a breach?	Yes: 6%	No: 40%	Never knew I could do so: 50%	Sometimes: 4%
27	Will you assert that your behaviors regarding your personal passwords are healthy? (This means good complexity, length, storage, nonreuse, etc.)2	Extremely healthy: 8%	Somewhat healthy: 44%	Neutral: 17%	Not very healthy: 24%	Not healthy at all: 7%
28	What is your proficiency level with technology/computers?	Beginner: 10%	Intermediate: 49%	Advanced: 31%	Expert: 10%
29	What gender do you identify as?	Woman: 63%	Man: 31%	Nonbinary: 1%	Prefer not to say: 5%

Countries of residence of respondents
USA: 84%; **Nigeria:** 8%; **India:** 1%; **England:** 1%; **Undisclosed:** 6%

11.4   ANALYSIS OF RESULTS

The survey respondents are from diverse backgrounds with 44% being under the age of 25, 31% between the ages of 26-45, 23% over the age of 45, and 2% choosing not to reveal which age bracket they belonged to. Females made up 63% of our respondents, 31% were male, 1% identified as nonbinary, and 5% preferred not to disclose their gender. Respondents came from four continents: Africa, Asia, Europe, and North America. Technology/computer proficiency was equal at 10% for beginner and expert respondents. The intermediate respondents were 49% and 31% identified as advanced. Students made up 36% of the respondents, 4% did not disclose their affiliation while the rest were either retired, an entrepreneur, chef, worked in: social media, private firms, education, government, IT, engineering, finance, health care, food service, hospitality, and public service.

The respondents were polled about their password hygiene and behaviors at the beginning and toward the end of the survey. We found that approximately 65% of respondents believed their password selection practices were robust and that their password behaviors, such as length, complexity and storage processes were sufficient. Conversely, 35% of the respondents reported their password hygiene as being either not very healthy, not healthy at all, or chose to remain neutral. Toward the end of the survey, we polled the respondents again about the health status of their password hygiene and behavior. We observed a gradual shift in responses with 52% of the respondents who believed their password hygiene and behaviors were healthy, while 48% remained neutral, not very healthy and not healthy at all.

We observed that 80% of respondents answered "yes" to being aware of two-factor authentication (2FA), while 56% answered "yes" to being aware or familiar with multi-factor authentication (MFA). Furthermore, for the question *"Some services such as online banking requires users to sign up for MFA. In cases where it is not required, what is the likelihood you would enable it?"* showed that respondents favored the convenience side of the convenience vs. security argument. The results showed 53% of the respondents reported being likely to enable MFA, 38% were neutral and unlikely to do so while 10% reported not knowing what MFA meant nor its purpose. Having 2FA or MFA enabled across their personal accounts, 52% of the respondents reported as often having 2FA or MFA enabled across their personal accounts. Reporting as neutral were 34%, while 14% combined reported rarely enabling it or do not enable it at all.

Most of the respondents, 76%, report having passwords that have less than 16 characters, 26% have passwords that have 8 or fewer characters, and 58% reporting having passwords that have at most 10 characters. At 73%, the respondents with advanced technology/computer proficiency had passwords that contain 16 characters or less. The advanced respondents, 67% of them reported having passwords that contained at most 10 characters and 33% with passwords containing 8 characters or less. Most of the respondents who identified as having expert proficiency reported having passwords that contained less than 16 characters at 71%. The expert users also reported having passwords that contain at most 10 characters at 86%. The results do not show a clear distinction between levels of proficiency and password length. Most of the respondents, irrespective of proficiency levels, had short passwords. Higher proficiency levels did not mean longer passwords, but beginner proficiency levels equated to shorter passwords. The beginners at 79% had passwords that contained 16 characters or less and equally split at 50% for passwords that have 8 or fewer characters or at most 10 characters.

The respondents who were less than 18 years old had passwords that were less than 16 characters at 79%, passwords with 8 or less characters at 39%, and at 61%, passwords that had at most 10 characters. The 18–35 years old that have passwords with 16 characters or less amount to74%, those with passwords that contain 8 characters or less are 28%, and those with passwords that are at most 10 characters are at 72%. The age groups of 36-55 and those above 55 years old each have passwords that are 16 characters or less at 78% and 77%, respectively. For passwords that contain 8 characters or less, 39% and 31%, and finally, 61% and 69%, respectively, for passwords that contain at most 10 characters.

Analyzing the respondents' password storage behaviors revealed a variety of behaviors with 50% of the respondents exhibiting multiple storage behaviors and reliant on memorization. The respondents who only stored passwords using plain text files on an electronic device were reported at 11%, passwords handwritten in a book/paper or other stationary medium was reported at 18%. Those who only stored passwords in the cloud: Google, iCloud, etc. and those who only stored their passwords in browsers for use via autofill were 7% and 14%, respectively.

Only 6% of the respondents utilized a password verification service such as *https:// haveibeenpwned.com/Passwords* [25] to verify if their passwords or credentials were part of a breach. Half of the respondents did not know it was possible to verify if their passwords or credentials were part of a breach. Another 40% of the respondents reported that they do not utilize any password verification tool or service while 4% reported doing so sometimes. This leads us to conclude that a knowledge gap and a bit of indifference to password verification exists.

The respondents reported sharing passwords. 58% of the respondents who reported sharing passwords did so with passwords to digital streaming platforms, email, banking/financial, e-commerce (Amazon Walmart, etc.) and other services. When asked if the passwords to personal accounts that they own were shared, 58% shared their passwords even if it was seldom done while 42% reported never sharing their passwords.

While the results show that the respondents share passwords, it also shows that 86% of the respondents reuse passwords with 14% not reusing passwords. Reusing the same password for 10 or more accounts stands at 19%. The respondents at 78%, believe that the regularity of password change requirements influences password reuse. Another 16% believe that password change requirements rarely influence password reuse while 6% report that it does not influence password reuse. Interestingly, 95% of the respondents' state that they are likely to reuse a password if allowed to do so.

The results showed that 48% of the respondents reported facing challenges generating good passwords, with another 25% being neutral, and 28% who reported not facing a challenge generating passwords. We asked the respondents if *ji32k7au4a83* was considered a good password. No context was provided, and the respondents were to decide without apriori knowledge. The respondents who did not consider it a good password stood at 48%, while 31% considered it a good password, and 22% could not decide selected maybe as their choice. The use of personal information while generating passwords is evident in the results as 54% of the respondents report using personal information in their passwords. Additionally, we observed that 61% of respondents use English for their passwords, while 39% use non-English languages in their passwords.

We asked if the respondents used a password manager for password generation and 17% responded with rarely, 31% do so at least sometimes, and 53% do not use a password manager for password generation at all. For personal use, 33% of the respondents use a password manager to store their passwords. Another 20% do so sometimes, while 47% do not safeguard their passwords using password managers.

11.5 DISCUSSION – ANALYSIS OF STUDY OPINIONS

We can draw a conclusion from the survey results that a decline in confidence of the respondents' assertion about their password hygiene and behaviors was impacted by the several questions in between the first and second prompt about password hygiene and behaviors. We believe that the respondents became more informed about password hygiene and undesirable behaviors from the questions asked, hence, the change in their password health assertions.

We asked the respondents if *ji32k7au4a83* was considered a good password without any context provided. We should have asked the respondents why they made their selection of Yes/No/Maybe. Some may have concluded it is bad because it is not easy to remember, or its short (sub 16 characters) or it only contains two classes (a–z and 0–9), etc. The respondents who considered *ji32k7au4a83* a good password stands at 31%, 22% were undecided and 48% did not consider it a good password. At times like these during password creation or verification, the ability to verify the safety and strength of a password could be increased by using a password manager or verifying via sites such as *https://haveibeenpwned.com/Passwords* [25]. Respondents did not consider that seemingly random selections of characters as passwords could in fact result to common, dictionary words (in this case common Mandarin) or part of a breach as is the case with *ji32k7au4a83* as a password. The password *ji32k7au4a83* is part of several breaches and translates to *"my password"* when using a Taiwanese keyboard with the Zhuyin Fuhao [26] format hence, making it quite common and unsuitable. Although it may look random and fairly complex at face value to a native English speaker or user, it is incredibly common and therefore a bad password.

Creating a good password can be challenging but it is not a task in futility, password managers can store and create strong, long, random, breach-free passwords. Unfortunately, more than half of the respondents of the survey do not utilize one. Cost, not needing one, lack of trust for password managers, too many choices to select from for password managers, and learning how to use them are the main reasons the respondents report being prohibitive to their use. It eliminates memorization and can minimize the need to reuse a password. 48% of our respondents report facing challenges generating good passwords.

We can draw some conclusions that the challenges from creating a good password could influence reuse. At 86%, we can conclude with high confidence that the respondents reuse passwords. A respondent reported that they reuse the same password "quite possibly 25+" times for accounts. Another respondent reported that their passwords are "all versions of my school password." Convenience, simplicity, easy to remember, and laziness are almost unanimous as the respondents' reasons for password reuse. Humans typically do not do well with the memorization of complex, random passwords. Due to this human limitation, reuse tends to flourish. Users may find a "good" password and stick to it, "modify" it a bit, or memorize it and therefore use it over and over. To avoid writing or storing passwords, users tend to include personal identifiable information since it is "easier" to remember. Unknowingly sometimes, this may avoid a "storage" or memorization challenge but introduce a vulnerable practice that creates a poor, undesirable and weak(er) password posture. Over half of the respondents reported having personal identifiable information contained in their passwords.

Half of the respondents stored their passwords with more than one medium such as in plain text files, in the Cloud, in browsers, or handwritten on some stationary medium. Only 7% reported using the cloud as their only means of password storage and 14% used only browsers for password storage. The largest solitary means of password storage was handwritten in books or on paper at 18%. Handwritten, plain text files, the Cloud may not particularly be bad but can be vulnerable. Storage of passwords in browsers is particularly not desired due to how vulnerable and dangerous it can be. Storage of passwords in a browser is convenient for users but also a popular target from bad actors via malware. We notice that the typical trends and behaviors have not changed as it pertains to storage with the respondents of the survey.

The results show that password sharing is still a widespread practice and behavior across age groups. Only the respondents above 55 years old showed a high resistance to sharing passwords at 75% irrespective of technology proficiency level. The other age groups were significantly lower than the above 55 years age group hence, showing a high rate of password sharing.

We analyzed the results of password lengths of the respondents across the different age groups in this order: <18 years, 18–35 years, 36–55 years, and > 55 years. We observed that, for passwords that contain at most 10 characters, those with less than 16 characters, and those with 16 characters or more averaged 66%, 77%, and 23%, respectively. This allows us to conclude that only a small number of respondents meet the vital characteristic of having a good password which is length. According to NIST, password length is the primary factor in characterizing password strength and composition [3].

11.6   CONCLUSIONS

We set out to investigate if the password behaviors of users differed from what has been observed in the past years in comparison to the current pervasive digital and social media era. Are password behaviors the same, different, or completely changed? We conducted a survey containing 35 questions to gain insights into the respondents' password behaviors.

We observed that the respondents of our survey reflected mostly similar password behaviors to the users from past years and those from before the pervasive social media era with smart devices. Our respondents just like their contemporaries in the years past, share their passwords, write their passwords in books or on paper, have short passwords (less than 8, 10, 16 characters), reuse their passwords across multiple accounts, encounter challenges creating good passwords, and experience knowledge gaps as it pertains to password hygiene.

For instance, the results show that the respondents view 2FA and MFA as different or separate in effect and in function as well. Respondents reported knowing the function and purpose of 2FA in relation to accounts but also did not know the same with MFA. This highlights a knowledge gap observed from the survey results. The dangers of poor password hygiene do not seem to be a deterrent to undesirable password behaviors. The respondents at 95%, acknowledged that they will reuse passwords if allowed to do so. Almost half of the respondents at 48% acknowledged to challenges in creating good passwords and half of all respondents were not aware of password verification tools or websites that could verify if a password was part of a breach or dictionary of common words. Almost 60% of the respondents reported sharing passwords to Digital Streaming Platforms, financial, email, and e-commerce sites. The level of technology/computer proficiency of the respondents had no impact on password length as expert users and beginners alike had passwords of less than 8 characters as well as those with at most 10 characters. Although we did not directly poll about the trust of password managers, the respondents reported with words such as: privacy, free versions are not convenient, lack of security, not safe, it can be exploited, hacked, or compromised, when asked about their password manager usage. Over half of the respondents do not utilize a password manager at all for password generation.

Our findings underscore the importance of password generation and management to our collective security posture. The results highlight the need for continuous education and reinforcement of good password habits, even among those who are technically proficient. Users need to adopt longer, but not necessarily more complex passwords, supplemented by, if not replaced by password managers. Avoiding undesirable password behaviors could significantly enhance and improve an individual's security posture.

11.7 SUGGESTIONS

- Password Management – Consider using a password manager for storage and creation of passwords.
- Password Verification – Considers using sites such as *https://haveibeenpw ned.com/Passwords* [25] or tools like your password manager to verify that your password is not weak, common, or part of a breach.
- Avoid sharing and reusing your password at all costs. Use a unique password for every account you use/own [20].

- Password Length – Is a vital characteristic of a good password as it mitigates against brute force attacks. Consider a minimum of 16 characters, the longer the better.
- Do not leave your accounts on devices unattended.
- Avoid storing your password and credentials in browsers for use with autofill.
- Education and awareness – Keep abreast of security training such as phishing, social engineering, and general security hygiene.
- Use MFA – This could be 2FA+ on all accounts when and if it is available.
- Password Creation sans Password Manager – Consider a *phrase* → *password* approach for password creation. For instance: This is the year of the 2024 Olympics in Paris → #..TitYr0tMM24OOOOiParis..#

ACKNOWLEDGMENTS

First, we thank Dr. Wayne Patterson for the opportunity to be collaborators for this book. We are very grateful! Thanks to Dr. Golden Richard III, Director of the Applied Cybersecurity Lab at Louisiana State University for making possible the collaboration between all the universities involved. Thank you to the members of the Applied Cybersecurity Lab of Louisiana State University specifically, Raphaela Mettig, Malana Fuentes, and Kaitlyn Smith for their contributions and reviews of the survey. A special mention to Kaitlyn Smith and Dajanae Vaughn for spreading the survey quickly.

REFERENCES

1. Merdenyan, Burak, and Helen Petrie. 2022. "Two studies of the perceptions of risk, benefits and likelihood of undertaking password management behaviours." *Behaviour & Information Technology* 41, no. 12: 2514–2527.
2. Gärdekrans, Rasmus, 2017. "Password Behaviour: A Study in Cultural and Gender Differences." Bachelor Degree Project, University of Skovde, Skovde, Sweden , 72 pp.
3. National Institute of Standards and Technology (NIST), 2024. [Online]. https://doi.org/10.6028/NIST.SP.800-63b
4. Reader's Digest, 2024. [Online]. www.rd.com/article/passwords-hackers-guess-first/
5. Komanduri, Saranga, Richard Shay, Patrick Gage Kelley, Michelle L. Mazurek, Lujo Bauer, Nicolas Christin, Lorrie Faith Cranor, and Serge Egelman, 2011. "Of passwords and people: Measuring the effect of password-composition policies." In *Proceedings of the SIGCHI Conference on Human Factors in Computing Systems*, pp. 2595–2604..
6. Tan, Joshua, Lujo Bauer, Nicolas Christin, and Lorrie Faith Cranor, 2020. "Practical recommendations for stronger, more usable passwords combining minimum-strength, minimum-length, and blocklist requirements." In *Proceedings of the 2020 ACM SIGSAC Conference on Computer and Communications Security*, pp. 1407–1426.
7. IBM, 2024. Cost of a Data Breach Report 2023 [Online]. www.ibm.com/security/data-breach
8. Impeva, 2024. Consumer Password Worst Practices. [Online]. www.imperva.com/docs/gated/WP_Consumer_Password_Worst_Practices.pdf
9. Crowdstrike, 2024. Brute Force Attacks. [Online]. www.crowdstrike.com/cybersecurity-101/brute-force-attacks/

10. 23andMe, 2024. Addressing Data Security Concerns – Action Plan [Online]. https://blog.23andme.com/articles/addressing-data-security-concerns

11. NordVPN, 2024. Colonial Pipeline Cyberattack: What Happened, and What We've Learned [Online]. https://nordvpn.com/blog/us-pipeline-hack/

12. Bloomberg, 2024. Hackers Breached Colonial Pipeline Using Compromised Password [Online]. www.bloomberg.com/news/articles/2021-06-04/hackers-breached-colonial-pipeline-using-compromised-password

13. Gaw, Shirley, and Edward W. Felten, 2006. "Password management strategies for online accounts." In *Proceedings of the Second Symposium on Usable Privacy and Security*, pp. 44–55.

14. Tam, Leona, Myron Glassman, and Mark Vandenwauver, 2010. "The psychology of password management: A tradeoff between security and convenience." *Behaviour & Information Technology* 29, no. 3: 233–244.

15. Moustafa, Ahmed A., Abubakar Bello, and Alana Maurushat, 2021. "The role of user behaviour in improving cyber security management." *Frontiers in Psychology* 12: 561011.

16. AT&T, 2024. AT&T Smart Home Manager. [Online]. www.att.com/internet/smart-home/

17. Woods, Naomi, and Mikko Siponen, 2024. "How memory anxiety can influence password security behavior." *Computers & Security* 137: 103589.

18. AT&T, 2024. AT&T Addresses Recent Data Set Released on the Dark Web. [Online]. https://about.att.com/story/2024/addressing-data-set-released-on-dark-web.html

19. Netflix, 2024. An Update on Sharing [Online]. https://about.netflix.com/en/news/an-update-on-sharing

20. Schneier Bruce, 2024. Choosing Secure Passwords [Online]. www.schneier.com/blog/archives/2014/03/choosing_secure_1.html

21. Cybernews, 2024. [Online]. https://cybernews.com/best-password-managers/most-common-passwords/

22. Gibson Research Corporation, 2024. How Big is Your Haystack? [Online]. www.grc.com/haystack.htm

23. Ars Technica, 2024. 25-GPU cluster cracks every standard Windows password in <6 hours. [Online]. https://arstechnica.com/information-technology/2012/12/25-gpu-cluster-cracks-every-standard-windows-password-in-6-hours/

24. Hashcat, 2024. World's Fastest Password Cracker. [Online]. https://hashcat.net/hashcat/

25. Have I Been Pwned, 2024. Pwned Passwords. [Online]. https://haveibeenpwned.com/Passwords

26. The Verge, 2024. ji32k7au4a83 Is a Surprisingly Bad Password. [Online]. www.theverge.com/tldr/2019/3/5/18252150/bad-password-security-data-breach-taiwan-ji32k7au4a83-have-i-been-pwned

Index

For Product Safety Concerns and Information please contact our EU
representative GPSR@taylorandfrancis.com
Taylor & Francis Verlag GmbH, Kaufingerstraße 24, 80331 München, Germany